Porous Metals and Metallic Foams
MetFoam 2023

12th International Conference on Porous Metals and Metallic Foams (MetFoam 2023), held from July 5-7, 2023, Dresden Radebeul, Germany

Editor
Olaf Andersen

Peer review statement

All papers published in this volume of "Materials Research Proceedings" have been peer reviewed. The process of peer review was initiated and overseen by the above proceedings editors. All reviews were conducted by expert referees in accordance to Materials Research Forum LLC high standards.

Published under License by **Materials Research Forum LLC**
Millersville, PA 17551, USA

Published as part of the proceedings series
Materials Research Proceedings
Volume 39 (2024)

ISSN 2474-3941 (Print)
ISSN 2474-395X (Online)

ISBN 978-1-64490-308-7 (Print)
ISBN 978-1-64490-309-4 (eBook)

This book contains information obtained from authentic and highly regarded sources. Reasonable efforts have been made to publish reliable data and information, but the author and publisher cannot assume responsibility for the validity of all materials or the consequences of their use. The authors and publishers have attempted to trace the copyright holders of all material reproduced in this publication and apologize to copyright holders if permission to publish in this form has not been obtained. If any copyright material has not been acknowledged please write and let us know so we may rectify in any future reprint.

Distributed worldwide by

Materials Research Forum LLC
105 Springdale Lane
Millersville, PA 17551
USA
https://www.mrforum.com

Manufactured in the United State of America
10 9 8 7 6 5 4 3 2 1

Table of Contents

Keywords

Editorial

Dear reader,

of the proceedings of the 12th International Conference on Porous Metals and Metallic Foams.

The MetFoam conference series is the largest cross-disciplinary, international technical meeting focused exclusively on the production methods, properties and applications of functional and lightweight porous/cellular metallic materials. The MetFoam conference series is kept and overseen by a steering committee composed of international experts. The committee confers regularly and appoints a General Chair for organizing each conference. MetFoam has been held so far in Bremen (1999, 2001) and Berlin (2003), Germany; Kyoto, Japan (2005); Montreal, Canada (2007); Bratislava, Slovakia (2009); Busan, Korea (2011); Raleigh, NC, USA (2013); Barcelona, Spain (2015); Nanjing, China (2017), and Dearborn, MI, USA (2019).

After two postponements due to the pandemic, it was finally possible to stage the 12th issue of the MetFoam conference series which was held from July 5-7, 2023, on the beautiful premises of the Radisson Blu Park Hotel & Conference Center in Dresden Radebeul, Germany. The total number of participants at MetFoam 2023 was 54. The analysis of the participant data showed that the largest group of participants came from Germany (18), followed by the USA (9) and South Korea (5). A total of 7 papers were submitted, which were peer reviewed by several steering committee members and Fraunhofer IFAM staff and are now published in these proceedings. I would like to thank all authors and reviewers for their contribution to this proceedings book as well as the Deutsche Forschungsgemeinschaft DFG for their substantial financial contribution to the conference, which made possible the successful restart of the MetFoam conference series.

Yours sincerely,
Dr.-Ing. Olaf Andersen
General Chair of MetFoam 2023

Committees

General Chair 2023
Olaf Andersen, Fraunhofer IFAM Dresden, Germany

Steering Committee 2023
John Banhart, Technical University of Berlin, Germany
David Dunand, Northwestern University, USA
Nihad Dukhan, University of Detroit Mercy, USA
Kiju Kang, Chonnam National University, South Korea
Louis-Philippe Lefebvre, National Research Council, Canada
Hideo Nakajima, Iwatani R&D Center, Japan
Afsaneh Rabiei, North Carolina State University, USA
Huiping Tang, Zhejiang University City College, China
Dong-hui Yang, Hohai University, China

Porous Metals and Metallic Foams: MetFoam 2023
Materials Research Proceedings 39 (2024) 1-8

Materials Research Forum LLC
https://doi.org/10.21741/9781644903094-1

Surface Emissivity Effect on the Performance of Composite Metal Foam against Torch Fire Environment

Nigel Thomas Amoafo-Yeboah[1] and Afsaneh Rabiei[1,a*]

[1]Mechanical and Aerospace Engineering, 1840 Entrepreneur Drive, Raleigh NC 27606, United States of America

[a]arabiei@ncsu.edu

Keywords: Surface Emissivity, Composites, Torch Fire, Metal Foams

Abstract. According to the US Department of Transportation (DOT), there are millions of liters of hazardous materials (HAZMATS) transported each year via railroad. This has translated to stringent safety measures taken to alleviate the effects of accidents involving tank cars carrying these HAZMATs. One of such measures is in the creation of the thermal protection system of tank cars in which the tank car must have sufficient thermal resistance when subjected to a simulated pool fire for 100 mins and a torch fire for 30 mins without its back plate temperature exceeding 427 °C at any point of time. This requires a suitable material as a thermal blanket and insulation in tank car lining. Steel-steel composite metal foam (S-S CMF) is a novel metal foam with unique properties of high strength to density ratio, lightweight, and high energy absorption. It consists of metallic hollow spheres that are closely packed within a metal matrix. The large percentage of air within the hollow spheres provide both lightweight and insulating effects for CMF. S-S CMF is being investigated using the standard torch fire test requirement to determine its suitability as a material for tank car thermal protection. This is accomplished by developing a numerical model using the Fire dynamics simulator (FDS) as a form of validation for experimental work done. To properly evaluate this, there are various thermal properties of S-S CMF that need to be established for predicting CMF's thermal response. Surface emissivity has been a challenging property to evaluate and hence this study focuses on developing an experimental and numerical procedure in evaluating this property for composite materials such as CMF. Preliminary data shows an acceptable prediction of emissivity, which will be applied to the FDS model for the torch fire test.

1. Introduction

Safety in the transportation industry has been a major interest for the US Department of Transportation (DOT), specifically in transportation of hazardous materials (HAZMATS). Currently the DOT have developed the DOT117 tank car with the DOT 117R standard specifications for thermal protection [1]. This specification requires the tank car to have sufficient thermal resistance when subject to a simulated pool fire for 100 mins and a torch fire for 30 mins [2]. Steel-Steel composite metal foam (S-S CMF) is a composite material that exhibits superior thermal properties as proven in some research studies [3], [4]. This material is being investigated for its potential use in these tank cars, as a partial replacement for the current configuration where ceramic wool is sandwiched by thick and heavy layers of carbon steel. So far, S-S CMF has successfully passed a pool fire and a small-scale torch fire test by a large margin [3][5]. However, since the desired goal is to have a full-scale torch fire test, it became necessary to develop a numerical method to expand upon data obtained from the small-scale experimental work. This is achieved using a fire-driven flow computational software known as the Fire Dynamics Simulator (FDS). This attempt requires modeling the S-S CMF by having known physical and thermal properties such as density, thermal conductivity, emissivity, and specific heat capacity. Due to the diverse parameters affecting the surface roughness of complex surfaces of composite metal foam, its surface emissivity measurements tend to pose a challenge when being experimentally evaluated

Porous Metals and Metallic Foams: MetFoam 2023   Materials Research Forum LLC
Materials Research Proceedings 39 (2024) 1-8   https://doi.org/10.21741/9781644903094-1

or analytically extrapolated. This study seeks to establish an experimental method for deriving surface emissivity, that is applicable to composite materials.

2. Materials and processing

The S-S CMF samples were manufactured using stainless steel hollow spheres embedded in a 316L stainless steel matrix. The lost core technique was used to manufacture hollow steel spheres with an average outer diameter of 2 mm with wall thickness of about 100 μm [6]. The spheres were manufactured in Dresden, Germany, by Hollomet GmbH. The hollow steel spheres were shaken into a random-loose packing arrangement within a steel mold surrounded by 316L stainless steel powder from North American Höganas, with an average particle size of 44 μm. The mold was then heated in a vacuum hot press and allowed to cool under a high vacuum to room temperature. The elemental composition of these samples is shown in Table 1.

After this, samples were then extracted, cleaned, and ground to create a flat surface and uniform thickness. Samples of cylindrical shape with dimensions of diameter 25.59 mm and height 25.26 mm were produced for this surface emissivity measurement. The samples were further ground on 180, 600 and 1200 grit sand paper.

Table 1. Chemical composition of S-S CMF components by weight percent

Material	Chemical composition (weight percent)						
	Fe	C	Mn	Si	Cr	Ni	Mo
2mm diameter Steel Spheres	Balance	0.68	0.13	0.82	16.11	11.53	2.34
316L Steel Matrix	Balance	0.03	2.00	1.00	16.00-18.00	10.00-14.00	2.00-3.00

3. Experimental procedure

3.1 Emissivity measurements

This section describes the test set up as well as the experimental procedure in determining emissivity of the S-S CMF samples. As a composite material, the S-S CMF is made up of spheres and matrix in between, and hence in determining the global emissivity, the emissivity's of the spheres and matrix were found separately and then put together using the rule of mixtures.

3.1.1 Test set up and measurement procedure for CMF matrix emissivity

The set up was developed in accordance with ASTM standard E1933—14 Standard Practice for Measuring and Compensating for Emissivity using Infrared Imaging radiometers [7]. For these tests the contact thermometer method was used. For emissivity measurements a FLIR E40 infrared radiometer camera was used. A Thermo Scientific Thermolyne HPA2235MQ analog hot plate was used as a primary heat source, with the CMF sample placed in a ceramic insulation block surrounding the perimeters of the cylinder, leaving the top surface exposed towards the IR camera and the bottom surface in direct contact with the hotplate. Two type R thermocouples were placed on the matrix surface of the sample for temperature measurements and were connected to a national instruments cDAQ 9171 data acquisition unit. The national instruments cDAQ was connected to a computer and LabVIEW software for the analysis of temperature data. An enclosure with a black interior was built to house the hotplate and sample, to prevent external reflections that could affect accuracy of the data acquisition. Figure 1 shows the test set up for the emissivity measurements. The set up was calibrated, by measuring emissivity of a flat stainless-steel sample from room temperature to 200 °C, resulting in data consistent with known emissivity values for stainless-steel.

Figure 1. (a) Overview of test set up (b) Image showing ceramic insulation on hot plate (c) CMF sample with thermocouples (d) IR camera reading emissivity data

To record emissivity data for the CMF matrix, the sample was heated from room temperature to 200 °C in increments of 50 °C. At each temperature increment, the FLIR IR camera was positioned to measure the temperature of the CMF matrix surface, and emissivity value on the IR camera adjusted till the camera temperature readings matched that off the temperature from the thermocouples. This adjusted emissivity value then becomes the emissivity of the CMF matrix at that temperature. This procedure was repeated for the various grit sizes of 180, 600 and 1200 to investigate the effect of surface roughness on emissivity. Results obtained from these measurements are discussed in the results section.

3.1.2 Measurement procedure for sphere emissivity

The second part of the procedure involves establishing a method for determining the emissivity of the spheres. Since the spheres are randomly packed in the matrix, and are cut at various sections, spheres are seen on the surface with varying depths, however this method does not take into consideration those different depths of spheres that are exposed on the top surface. However, an indirect method is adapted to measure the sphere emissivity's. This involves measuring the surface roughness of the spheres and then producing this roughness on a regular 316L stainless steel sample using the sand blasting method. These roughness measurements were done using a Keyence VKx1100 optical profilometer. This device scans the sample by shooting a laser to detect the height (z) information across the x and y planes of the sample surface and gives a roughness (Ra) value for the profile. This was done multiple times across the x and y plane to give an average Ra value. After measurements were done, the average Ra valued obtained for the sphere surface was 5.4 μm, whilst that of the sandblasted sample was 6.8 μm, translating to close enough values for emissivity measurements. Once matching the surface roughness of the interior surfaces of spheres and the sandblasted sample were successful, emissivity measurements were conducted on the sand blasted sample at temperatures between room temperature and 250 °C. Table 2 shows raw data obtained for emissivity obtained after measurement.

Table 2. Emissivity measurements obtained for Sand Blasted sample.

Emissivity measurement for sand blasted sample	
Temperature (°C)	Emissivity
50	0.8
100	0.8
150	0.75
200	0.63
250	0.58

3.1.3 Effects of sphere curvature on emissivity

Emissivity is affected by several factors, one of which is the angle of measurement which is depicted in figure 2. Since sphere emissivity's are being measured indirectly, it was needful to come up with a numerical formulation to take into consideration the effects of the spheres interior surface curvatures on the global sphere emissivity. This was achieved by generating an equation based on a compilation of available literature data on the effect of measurement angle on 316L stainless steel [8][9][10]. These literature sources derived similar equations, and hence were numerically combined with data from the sandblasted 316L sample as a predictive method. This was done by equating the y intercept of the equation to the emissivity of the sand blasted sample at room temperature, to give emissivity for the interior surfaces of CMF spheres at various angles. The equation used is shown below:

$$y = 1E\text{-}11x^6 - 2E\text{-}09x^5 + 2E\text{-}07x^4 - 1E\text{-}05x^3 + 0.0002x^2 - 0.0027x + 0.1697 \qquad (1) [9]$$

where y = emissivity of sphere and x = measurement angle

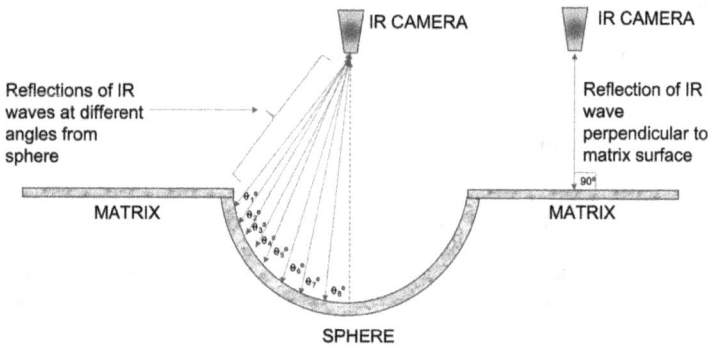

Figure 2. Schematic showing effect of curvature on emissivity data.

4. Results and discussions

4.1 Matrix emissivity

Figure 3 shows measured emissivity of the matrix of the CMF samples at various temperature points and varied surface roughness. It can be observed that the surface emissivity values are decreased with increasing temperature. Also, there is a considerable reduction in the surface emissivity values as the surface roughness decreases and this is better seen with an increase in temperature. This confirms that the surface roughness directly impacts emissivity.

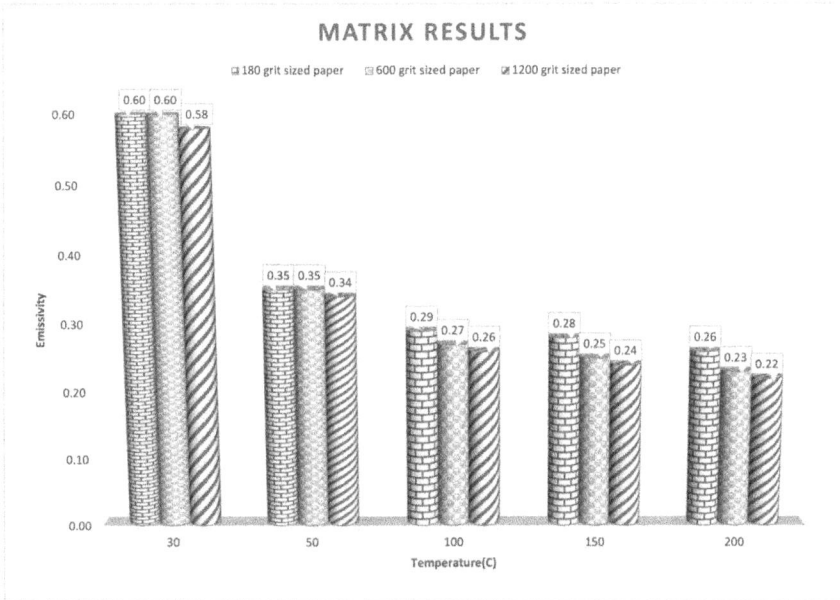

Figure 3. CMF matrix emissivity at various surface roughness and temperatures.

4.2 Sphere emissivity results

Figure 4 shows emissivity results obtained from the sand blasted sample and combined with the effects of spheres angle. It is seen that due to the spheres having a rougher surface profile than that of the matrix, emissivity values obtained here are higher, ranging from 0.5 to 0.99. The angle effects on the sphere emissivity were averaged out and resulting values shown in Table 3. These values become representative of the emissivity of the CMF spheres, which will be applied in the next section for the rule of mixtures.

Porous Metals and Metallic Foams: MetFoam 2023   Materials Research Forum LLC
Materials Research Proceedings 39 (2024) 1-8   https://doi.org/10.21741/9781644903094-1

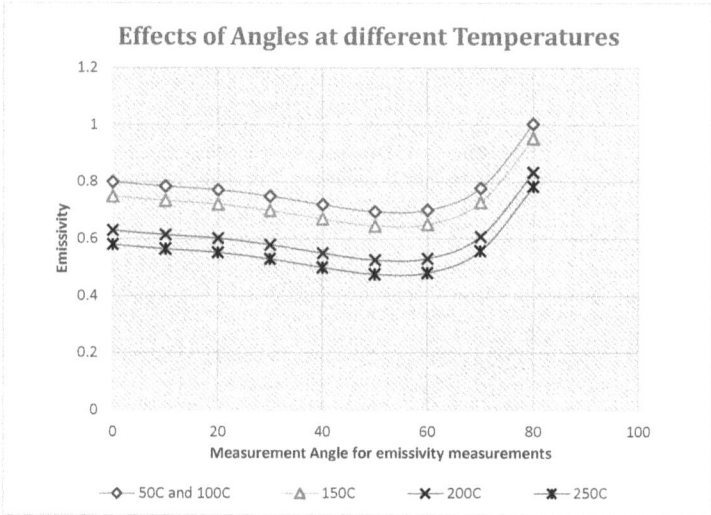

Figure 4. Effects of angle on emissivity of spheres.

Table 3. Averaged emissivity values of CMF spheres.

Average Emissivity of CMF Spheres	
Temperature (°C)	Emissivity
50	0.778
101	0.778
152	0.728
203	0.608
248	0.558

4.3 Rule of mixtures

The lower bound and upper bounds rule of mixtures was applied to the S-S CMF to predict the global surface emissivity of the material based on the emissivity of its components (hollow spheres and the matrix) using the following formula:

$$Upper\ Bound: \varepsilon_{cmf} = A_{spheres} \cdot \varepsilon_{spheres} + A_{matrix} \cdot \varepsilon_{matrix} \tag{2}$$

$$Lower\ Bound: \varepsilon_{cmf} = \frac{1}{[(A_{matrix}/\varepsilon_{matrix}) + (A_{sphere}/\varepsilon_{spheres}]} \tag{3}$$

where ε = emissivity, A = area.

Using a scanning electron microscope (SEM) image of the CMF sample surface and an image J software the %area of sphere and matrix on a 2 mm sphere CMF sample is estimated (Figure 5(b)).

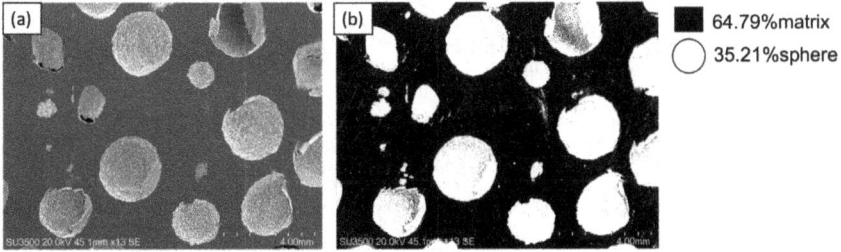

Figure 5. (a) SEM image of CMF sample (b)Image J used to determine %area of spheres and matrix.

After the image J data for area measurements is incorporated into the rule of mixture equations, data was extrapolated to 1200 °C to predict the global surface emissivity of S-S CMF at higher temperatures. Figure 6 shows data obtained. To better understand the graph, "UB" and "LB" represent data for upper-bound and lower-bound emissivity, respectively based on the direct experimental results and shown with scattered graphs."EXT" represents extrapolated data from 200C to 1200C shown by line graphs. The experimental results were collected from CMF sample surfaces ground at 180, 600 and 1200 grit sizes tested at various temperatures. As can be seen, the emissivity of CMF is directly affected by the roughness of the CMF surface. The smoother the surface is the lower the emissivity value hence data for 1200 grit sized samples show the lowest emissivity values, followed by 600 grit and 180 grit. Emissivity of CMF is also impacted by testing temperature, with higher temperatures showing higher emissivity.

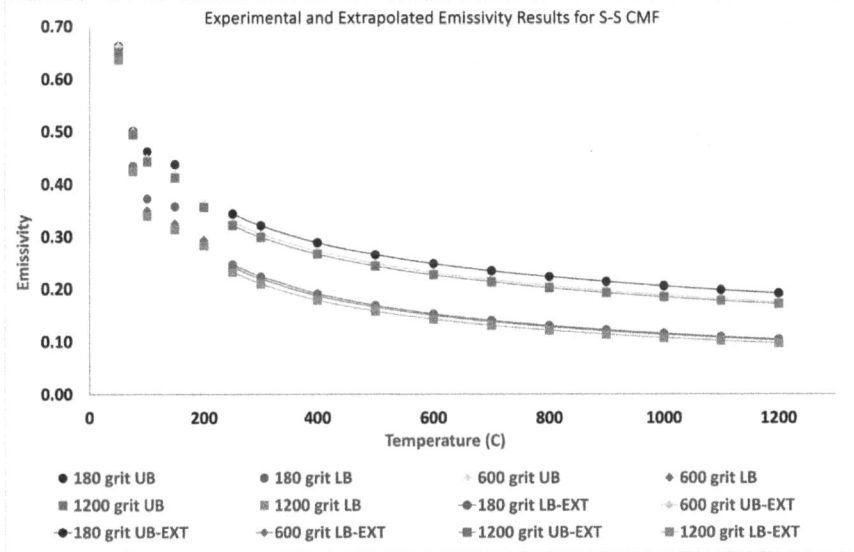

Figure 6. Lower Bound (LB) and Upper Bound(UB) experimental and extrapolated (EXT) surface emissivity values of 2mm sphere S-S CMF for 180, 600 and 1200 grit size at various temperatures.

Conclusion

A systemic approach for the measurement of the surface emissivity of composite materials in general and composite metal foams in particular was successfully developed, and data derived will be applied to a numerical model for a full-scale torch fire test model. Data generated showed that the method developed for finding surface emissivity can predict the surface emissivity of complex composite materials with an easy approach of rule of mixture.

Acknowledgements

This study was supported by the Department of Transportation (DOT) Pipeline and Hazardous Materials Safety Administration (PHMSA) under project number #PH957-20-0075.

References

[1] United Sates Department of Transportation, "Tank Car Specifications and Terms," *Bureau of Transportation Statistics*, Apr. 18, 2018. https://www.bts.gov/surveys/annual-tank-car-facility-survey/tank-car-specifications-terms

[2] Legal Information Institute, Cornell Law School, "49 CFR Part 179- Specifications For Tank Cars." https://www.law.cornell.edu/cfr/text/49/part-179

[3] A. Rabiei, K. Karimpour, D. Basu, and M. Janssens, "Steel-steel composite metal foam in simulated pool fire testing," *International Journal of Thermal Sciences*, vol. 153, p. 106336, Jul. 2020. https://doi.org/10.1016/j.ijthermalsci.2020.106336

[4] S. Chen, J. Marx, and A. Rabiei, "Experimental and computational studies on the thermal behavior and fire retardant properties of composite metal foams," *International Journal of Thermal Sciences*, vol. 106, pp. 70–79, Aug. 2016. https://doi.org/10.1016/j.ijthermalsci.2016.03.005

[5] Nigel Amoafo-Yeboah and R. Afsaneh, "Thermal Response of Steel-Steel Composite Metal Foams under Small-Scale Torch-Fire Conditions," *Advanced Engineering Materials*, May 2023. https://doi.org/10.1002/adem.202300217

[6] B. P. Neville and A. Rabiei, "Composite metal foams processed through powder metallurgy," *Materials & Design*, vol. 29, no. 2, pp. 388–396, Jan. 2008. https://doi.org/10.1016/j.matdes.2007.01.026

[7] "ASTM E1933-14(2018) Standard Practice for Measuring and Compensating for Emissivity Using Infrared Imaging Radiometers."

[8] L. Barker, "Influence of oxidation and emissivity for metallic alloys space debris during their atmospheric entry," in *7th Eurpean Conference on space debris*, Germany, Apr. 2017.

[9] D. Atasi, "Temperature and angle dependent emissivity and thermal shock resistance off the W/WAlN/WAlON/Al2O3 based spectrally selective absorber," *Applied energy materials*, vol. 2, pp. 5557–5567, 2019. https://doi.org/10.1021/acsaem.9b00743

[10] L. Yang and X. Xin-Lin, "Tomography-based analysis of apparent directional spectral emissivity of high porosity nickel foams," *International Journal of Heat and Mass Transfer*.

Porous Metals and Metallic Foams: MetFoam 2023
Materials Research Proceedings 39 (2024) 9-16

Materials Research Forum LLC
https://doi.org/10.21741/9781644903094-2

Characterization of 316L Stainless Steel Composite Metal Foam Joined by Solid-State Welding Technique

John M. Cance[1] and Afsaneh Rabiei[1, a]

[1] Department of Mechanical and Aerospace Engineering, NCSU, Raleigh, NC 27695, United States

[a]arabiei@ncsu.edu

Keywords: Composite Metal Foam, Solid-State Welding, Microstructure, Mechanical Properties

Abstract. In previous studies, composite metal foams (CMF) have shown exemplary mechanical performance under impact which has made them prime candidates for protection of transported passengers and cargo. [1] Materials utilized in such applications often require joining to form structures and geometries that are far more complex or impossible to produce in an as-manufactured state. Welding methods are popular in the joining of metals with solid-state welding processes such as induction welding being of particular interest in the studies to be discussed. In this study, various thicknesses of 316L stainless steel CMF are manufactured through powder metallurgy technique and welded using Induction Welding. The mechanical properties of the weldments were studied through uniaxial tensile tests while microstructural characterization of the weldment within the joint interface and heat-affected zone (HAZ) are evaluated using scanning electron microscopy. The combination of these evaluations grant insight on the effects of various weld parameters (e.g., welding temperature, workpiece thickness, flux, and welding environment) as well as the suitability and restrictions of induction welding in the joining of 316L Stainless Steel CMF.

1. Introduction

The industrialized structure of first-world countries has steadily necessitated a heightened volume of resources to be transported domestically through roads and railways in recent decades. Railways in particular account for almost half of the freight transported within the United States with unparalleled fuel efficiency. However, roughly 75% of this content consists of hazardous materials (HAZMAT) which, while necessary in forms such as fuels and chemicals, can harbor catastrophic results when introduced to the environment through tank car derailment and subsequent rupture. [2,3] One such scenario came to fruition in 2013 when 72 DOT-111 tank cars containing crude oil derailed in Lac-Mégantic, Quebec, Canada, resulting in fires that claimed 47 lives and destroyed much of the local infrastructure. [4] This tragedy resulted in the development of DOT-117 and DOT-117R standard tank cars, bolstered for HAZMAT transport by a layer of fire-retardant insulation as well as welded external jackets and head shields of TC-128B steel to dampen impact in the event of derailment. However, advancement is still pursued in this field to further improve durability of tank cars transporting HAZMAT and subvert potential tragedies caused by loss of lading.

Metal foams are unique family of materials characterized by their low-density porous structures and high strength-to-weight ratio, allowing impressive degrees of thermal insulation and impact energy absorption. [5] This performance has led to metal foams gaining preference in the automotive and aerospace industries to further improve user safety in the event of impact while simultaneously facilitating development of lightweight components. However, conventional metal foams are often mechanically unpredictable, due to their nonuniform porosity sizes, profiles, and distribution within the product. Under compressive loading, this can result in formation of collapse bands as larger cells implode, leading to a premature failure of the material with low energy absorption capacity. [5,6] Steel-steel composite metal foam (S-S CMF) has been shown to circumvent this effect by forgoing the traditionally foamed closed cells in favor of similar hollow 316L steel spheres surrounded by a sintered 316L steel matrix. This allows the cells to distribute compressive load with relative homogeneity while bonding between the matrix and outer sphere walls provides additional support, resulting in superior mechanical performance compared to conventional metal foams.[1]

Construction of large structures such as plane fuselages and railcars is heavily reliant on welding of constituent materials, such as metal sheets and panels. Despite the joining of bulk metals being widely explored, unique concerns are presented when determining suitable processes for welding porous metals. The most pressing one being preservation of the cellular structure, and its inherent benefits, without sacrificing integrity of the welded joint. Because of this, fusion welding methods are often overlooked as their propensity to liquify the bonded interface results in a weld more akin to a bulk material upon solidification, exacerbating the already present concern of inhomogeneous behavior. Conversely, solid-state welding methods have relatively minimal effects on the base material, substituting the intensity of a direct arc/flame in favor of induction or frictional heating combined with external pressure to encourage the mating of the bonding surfaces. Induction welding is of particular interest in this study. Induction coils were used to induce Eddy currents within the workpiece and set bonding temperatures were achieved through resistance between the induced current and workpieces. This results in a solid weld while minimizing heat effects on the material, making induction welding a suitable option for joining S-S CMF panels. In this study, the properties of S-S CMF panels of various thicknesses joined using induction welding will be presented and the benefits and limitations of this solid-state joining will be discussed.

2. Experimental

2.1 Materials and processing

The S-S CMF panels welded in this study were manufactured through the powder metallurgy (PM) method established in previous publications [7]. The 316 stainless steel hollow spheres with average outer diameters of 2mm and wall thickness of 0.1mm were used (produced by Hollomet GmbH located in Dresden, Germany). Additionally, matrix material was comprised of 316L stainless steel powder atomized by North American Höganäs High Alloys LLC. The resulting S-S CMF panels ranged in density from 2.5 to 3.3g/cc. The sintering process was followed by sectioning of the panels into rectangular workpieces using waterjet cutting. These reduced panels were then brought to final dimensions through facing and end milling. Final thicknesses were modified within a range of 10.2mm to 28.32mm to observe the influence of panel thickness on penetration of Eddy current during the induction welding process. In preparation for welding, intended bonding surfaces underwent additional milling and grinding to provide an even finish.

All induction welding of the S-S CMF panels was conducted by Advanced Materials Manufacturing (AMM), a startup company based in Raleigh, NC. An ECO-LINE MFG 100 and ECO-LINE MFG 500 high-frequency power sources from Eldec was used for this purpose. Two induction coil designs were used for both groups, interchanged based on panel thickness. Both designs possessed a C-shaped profile outfitted with rectangular ferrotrons along the length to facilitate a more uniform heating profile on the top and bottom of the welding panels, as well as along the interface length. Panels of S-S CMF were paired for joining based on corresponding thicknesses to avoid significant distortion at the weld and achieve consistent bonding. Both coils featured even distributions of ferrotrons along the top and bottom, with Coil #1 possessing a 20-27mm wide opening to accommodate thinner panels while Coil #2 enveloped thicker panels with a 40mm gap. Overall, the welding procedure was relatively similar throughout the selected pairs

Porous Metals and Metallic Foams: MetFoam 2023
Materials Research Proceedings 39 (2024) 9-16

Materials Research Forum LLC
https://doi.org/10.21741/9781644903094-2

of CMF panels (shown in Table 1). Each set was fastened within a vise, maintaining contact between the workpieces as they were heated by the induction coil, before cooling and removal. Welding temperatures were determined for each sample set using a FLIR infrared camera directed at the joint interface. This general arrangement of the induction welding process is shown below in Fig. 1, where the heated joint between CMF panels is highlighted in red. Despite these similarities, group 1 weld surface preparation was conducted using a mixture of water and Royal Tiger flux applied to the mating surfaces prior to welding. In this stage, all samples were welded in air. Conversely, group 2 omitted the water base flux to eliminate some oxidation issues. Additional runs were conducted in argon to minimize oxidation. These combined efforts resulted in welded S-S CMF panels with solid contact along the bonded interface and a surface relatively free of visible defects, as Shown in Fig. 2.

Table 1. Induction welded panels arranged by applied welding parameters.

Group	Thickness [mm]	Weld Temp [°C]	Environment	Flux
	10.2	1220	Air	Yes
1	15.2	969	Air	Yes
	17.7	1100	Air	Yes
	22.5	850	Air	Yes
	21.2	1050	Argon	No
2	26.39	1100	Air	No
	28.32	1070	Argon	No

Fig. 1. Recreation of the induction welding setup consisting of A) the FLIR IR camera, B) vise, C) abutted S-S CMF panels, and D) an induction coil outfitted with ferrotrons, displayed as solid dark grey regions.

Porous Metals and Metallic Foams: MetFoam 2023   Materials Research Forum LLC
Materials Research Proceedings 39 (2024) 9-16   https://doi.org/10.21741/9781644903094-2

Induction welding was followed by face milling to remove surface features that may affect tensile performance at the weld seam. Tensile specimens of the induction welded panels were designed based on AWS B4.0:2016 standards for transverse rectangular tensile test and extracted from the welded plates. Due to the varied sizing of welded panels produced, scaling was required to accomodate 2 to 3 tensile specimens per weld, allowing thorough mechanical characterization along the seam. Test specimens for group 1 were extracted through waterjet cutting then ground along the cut surfaces to remove any resulting imperfections, while group 2 panels were sectioned using a band saw and brought to their final profiles through end milling. A schematic of the tensile specimen arrangements along a pair of welded panels is shown below in Fig. 3, where the weld line is represented by a red line.

Fig. 2. Induction welded panels of steel-steel composite metal foam with an in-tact porous structure and surface free of defects. Some dark regions near the weld line indicate some degree of oxidation. Image courtesy of Advance Materials Manufacturing, LLC.

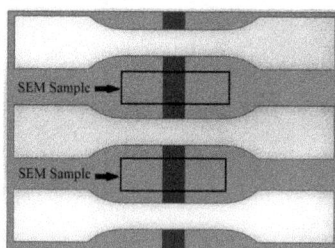

Fig. 3. Orientation of 3 tensile specimens along a pair of welded panels.

Specimens for microstructural analysis were sourced from the leftover material between the dog bone tensile specimen extraction sites then ground and polished using 180, 320, 600, 800, and 1200 grit Buehler SiC papers followed by 3 and 1μm Buehler MetaDi diamond suspensions, respectively. Both grinding and polishing processes took place on a Buehler AutoMet Powerhead 2 grinding and polishing station, with each grit being followed by ultrasonic cleaning in water then acetone before moving to the next grit to prevent cross contaminations. The welding area and its heat affected zone's microstructure was evaluated using a Scanning Electron Microscope (SEM).

Porous Metals and Metallic Foams: MetFoam 2023   Materials Research Forum LLC
Materials Research Proceedings 39 (2024) 9-16   https://doi.org/10.21741/9781644903094-2

Imaging was conducted using a Hitachi SU3500 SEM equipped with Electron Dispersive Spectroscopy (EDS) and Electron Backscatter Diffraction (EBSD).

2.2 Tensile testing

Tensile tests were executed in accordance with ASTM E8/E8M at a loading rate of 0.1mm/min, using a Q-Test Electromechanical Universal Testing Machine equipped with pneumatic grips and 20kip load capacity. Specimens failing at the weld line were set aside as outliers due to this type of failure representing an unsuitable bond. In turn, outliers were excluded from the data sets presented in this study.

3. Results and discussion

3.1 Preliminary structural observation

Preliminary SEM imaging of a cross-sectional S-S CMF strip extracted from the induction welded panels provided confirmation that the process was successful in joining the matrix between panels while effectively preserving the hollow spheres, with little distortion near the bonded interface. A broad view of a weld interface can be seen in Fig. 4 A) and B), showing a clean bond with only slight filling of the previously cut spheres being due to the forces imparted by thermal expansion while fixed in a vise during welding. The accompanying views in Fig. 4 C) and D) highlight higher magnification views of the fused matrix. The progressively higher magnification views of Fig. 4 are denoted sequentially by white squares showing their locations.

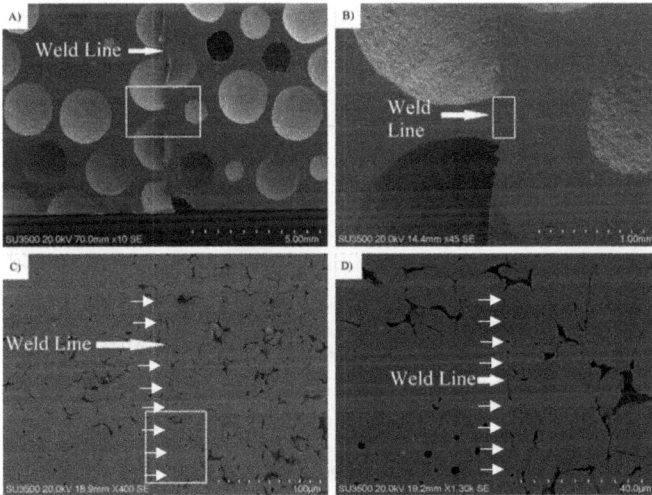

Fig. 4. A) A broad SEM image of the induction welded interface along S-S CMF panel, B) a magnified view highlighting a region of the weld, C) a further magnified view of the seamlessly welded matrix, and D) a final magnified image of the matrix, showing a near-seamless joint.

Rust was observed to be abundant throughout outlying tensile specimens produced in group 1, while being almost absent from group 2 specimens. This suggests the use of an argon welding environment fulfilled its desired role, as well as a possibility that the water-based flux used in group 1 may have contributed to oxidation. However, this effect requires further scrutiny to determine whether the flux is in fact problematic, or oxidation had alternatively resulted from sensitization of the heat affected zone (HAZ) due to precipitation of chromium carbide (Cr_3C_2). This effect is being studied and results will be reposted in future publication.

3.2 Mechanical behavior

Tensile tests revealed a wide range of performances across both sample groups, lending to the suspected impact of welding parameters on mechanical performance. Panel thickness and maximum temperature achieved during the welding process are the most important variables. Heat penetration is a documented limitation of induction welding [8], heavily influenced by phenomena known as skin effects. This occurs when the heating profile of a conductive material becomes distorted as induced Eddy current concentrates at the surface closest to the induction coil, resulting in a conical heating profile through the cross-section that narrows further from the coil. Depth of weld penetration in this case is inversely proportional to the welding frequency and magnetic permeability of the workpiece, rendering skin effects surmountable with iterative adjustment of the process parameters. [8] Implementation of the C-shaped coils was intended to prevent this, heating the workpiece from both top and bottom to optimize penetration of the heating profile, and appears to have been successful as obtained with the bonding of the 22.5mm thick S-S CMF panels (as shown in Table 1). However, Fig. 5 shows a steady decline in average ultimate tensile strength (UTS) of welds with increasing panel thickness., E1/E2, a 10.2mm thick panel, displayed an UTS of 63 MPa while 20 MPa was obtained with thicknesses approaching 25mm.. This trend suggests that thickness could be a limitation when induction welding S-S CMF.

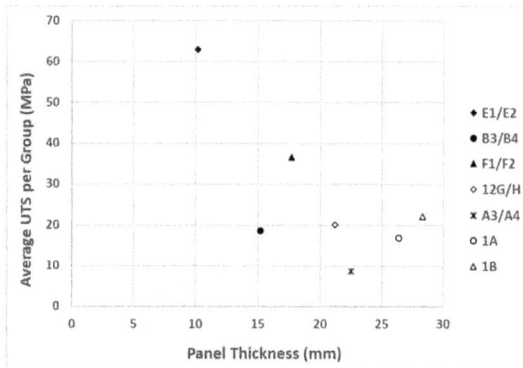

Fig. 5. Average ultimate tensile strength plotted against thickness of S-S CMF panels at the time of welding.

Contrary to the effect of thickness, rise in welding temperature appears to be directly related to the mechanical performance, as shown in Fig. 6. This behavior is expected as further liquefaction and solidification often intensify the degree of bonding between metals. A similar effect is present in the induction welded samples of this study with the panels essentially undergoing localized sintering at the bond interface. The sintering temperature of 316L stainless steel is known to dwell between 1100°C and 1300°C . [9] This correlates well with the observed tensile behavior, showing peak performance of 63 MPa paired with a welding temperature of 1220°C. While the current influences of thickness and temperature offer promising insight to the limitations and optimization of induction welding of S-S CMF, research is ongoing to isolate further influential factors such as coil size, welding power, preheating, and environment.

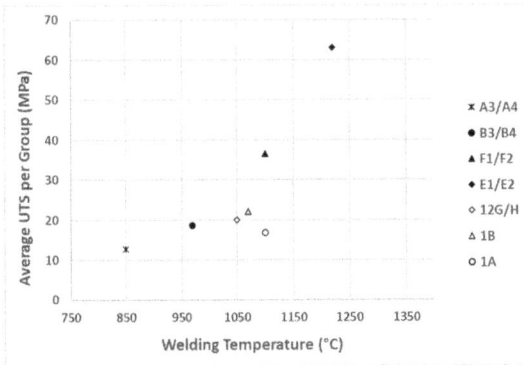

Fig. 6. Average ultimate tensile strength plotted against weld temperature of S-S CMF panels.

Conclusions

Observation and mechanical evaluation of S-S CMF panels joined through induction welding has granted a preliminary understanding of the viability of this process and its parameters. Broad SEM observations have shown that the process is overall successful in joining S-S CMF panels, with the matrix showing an indistinguishable interface and hollow steel spheres remaining nearly spherical with no distortion. Furthermore, welding temperatures comparable to sintering conditions were achieved, showing a clear benefit at higher temperatures, while panel thickness draws a clear limitation to the compatibility of induction welding of S-S CMF. Further investigation is ongoing to formulate decisive correlations between other process parameters and weld performance, though the present data appears promising.

Acknowledgements

The authors would like to express their gratitude to the United States Department of Transportation (DOT) for their financial support through the Pipeline and Hazardous Materials Safety Administration (PHMSA) award number PH957-20-0075. Additional gratitude is extended to Advanced Materials Manufacturing, LLC team of engineers and scientists to lead and conduct all induction welding works.

References

[1] J. Marx, A. Rabiei, Overview of Composite Metal Foams and Their Properties and Performance, Advanced Engineering Materials. 19 (2017) 1-13. https://doi.org/10.1002/adem.201600776

[2] Found on https://www.aar.org/wp-content/uploads/2021/02/AAR-Freight-Rail-Climate-Change-Fact-Sheet.pdf

[3] Found on https://tankcarresourcecenter.com/tankcar101/#1499693534177-d720d234-3972

[4] Found on https://www.tsb.gc.ca/eng/rapports-reports/rail/2013/r13d0054/r13d0054-r-es.html

[5] F. Garcìa-Moreno, Commercial Applications of Metal Foams: Their Properties and Production, Materials. 85 (2016) 1-27. https://doi.org/10.3390/ma9020085

[6] H.-P. Degischer, B. Kriszt, Handbook of Cellular Metals, WILEY-VCH Verlag GmbH, Weinheim, 2002. ISBN 3-527-60055-8

Porous Metals and Metallic Foams: MetFoam 2023
Materials Research Proceedings 39 (2024) 9-16

Materials Research Forum LLC
https://doi.org/10.21741/9781644903094-2

[7] B.P. Neville, A. Rabiei, Composite metal foams processed through powder metallurgy, Materials and Design 29 (2008) 388-396. https://doi.org/10.1016/j.matdes.2007.01.026

[8] T.J. Ahmed, D. Stavrov, H.E.N. Bersee, A. Beukers, Induction welding of thermoplastic composites - an overview, Composites: Part A. 37 (2006) 1638–1651. https://doi.org/10.1016/j.compositesa.2005.10.009

[9] P. Samal, J. Newkirk, Sintering of Stainless Steels, in: ASM Handbook Volume 7: Powder Metallurgy, AMS International, 2015, pp. 434-439. https://doi.org/10.31399/asm.hb.v07.9781627081757

Porous Metals and Metallic Foams: MetFoam 2023
Materials Research Proceedings 39 (2024) 17-24

Materials Research Forum LLC
https://doi.org/10.21741/9781644903094-3

Thermal Conductivity of Steel-Steel Composite Metal Foam through Computational Modeling

Zubin Chacko [1] and Afsaneh Rabiei [1,a] *

[1] Advanced Materials Research Laboratory (AMRL), Department of Mechanical and Aerospace Engineering, North Carolina State University, Raleigh, NC, 27606, USA

[a] arabiei@ncsu.edu

Keywords: Composite Metal Foams, Effective Thermal Diffusivity, Effective Thermal Conductivity, Computational Modeling, ANSYS Fluent, Thermal Insulation

Abstract. Thermal capabilities of Steel-Steel composite metal foam (CMF) against extremely high temperatures using computational methods have been investigated and contrasted with the characteristics of the base bulk steel materials. A physics-based three-dimensional model of CMF was constructed using Finite Element Analysis software for analyzing its thermal conductivity. The model built and analyzed in ANSYS Fluent was based on high temperature guarded-comparative longitudinal heat flow technique. ANSYS Fluent allows for the inclusion of air in the model, which is the main contributor to the low thermal conductivity of CMF compared to its constituent material. The model's viability was checked by comparing the computational and experimental results, which indicated approximately 2% deviation throughout the investigated temperature range. Excellent agreement between the experimental and computational model results shows that the CMF can be first modeled and analyzed using the proposed computational technique for the desired thermal insulation application before manufacturing. Based on the ratios of the matrix to the spheres and the thickness of the sphere walls, CMF can be tailored to the density requirements and then checked for its thermal performance using the model, thereby lowering the cost involved in its manufacturing and thermal characterization experiments.

Introduction

A type of metal foam called composite metal foam (CMF) is created using hollow metal spheres encased in a metallic matrix. The cell walls and the general integrity of the structure are strengthened by the metal matrix, which surrounds and fills the spaces between a loosely packed, random arrangement of hollow metal spheres. Casting [1] or powder metallurgy (PM) [2] are two methods used to make CMFs. Metal foams are renowned for being efficient and effective heat sinks due to the substantial surface area that the porosities in these materials provide [3]. A prior work [4] reported the specific heat capacity, effective thermal conductivity, and coefficient of thermal expansion of CMFs. However, it is crucial to note that only the temperature range between 300 and 600 °C was used to study the thermal conductivity values of S-S CMF. Identifying the fundamental thermal characteristics of S-S CMF at even higher temperatures, close to 1000 °C, is crucial for its application in the industry. CMF characteristics have been shown to vary with stainless steel spherical size and wall thickness and can be tailored for specific uses [5]. Therefore, it would be advantageous if crucial thermal characteristics, like the thermal conductivity of S-S CMF, could be anticipated using commercial modeling software. To develop the best S-S CMF for the intended application, sphere size, and wall thickness can be changed in computational solvers. Results can then be verified experimentally, thereby reducing the cost of testing.

This study uses a computational method to investigate the thermal conductivity of S-S CMF from ambient temperature to 1000 °C. First, to forecast the thermal conductivity of Composite metal foams with 2 mm steel hollow spheres and 316L stainless steel matrix [(2 mm sphere) S-S CMF] from 100 to 1000 °C, a physics based three-dimensional model was created utilizing the

Porous Metals and Metallic Foams: MetFoam 2023　　　　　　　Materials Research Forum LLC
Materials Research Proceedings 39 (2024) 17-24　　　　　https://doi.org/10.21741/9781644903094-3

high-temperature guarded-comparative longitudinal heat flow (GCH) technique. The results of the thermal conductivity data obtained through computational modeling were compared with experimental data to determine the model's viability.

Computational modeling

Model Description. First, the (2 mm sphere) S-S CMF is modeled based on the experimental sample used in the earlier work [4]. Spherical porosities were added and organized in a body-centered cubic (BCC) structure with inner diameters of 1.8 mm to mimic the geometry of the CMF with 2 mm stainless steel hollow spheres and 100 μm wall thickness. These porosities represent the amount of air inside the hollow stainless-steel spheres. Spacing was added between the porosities to represent the spherical wall thickness and achieve the average overall packing efficiency of S-S CMF, defined as 59% [2]. The sphere walls and matrix are assumed to be a single entity in the model to reduce the complexity of the model and speed up calculations. Finally, the matrix of the CMF has been regarded as a bulk material. Fig. 1(a) depicts the model of the CMF with the porosity configuration.

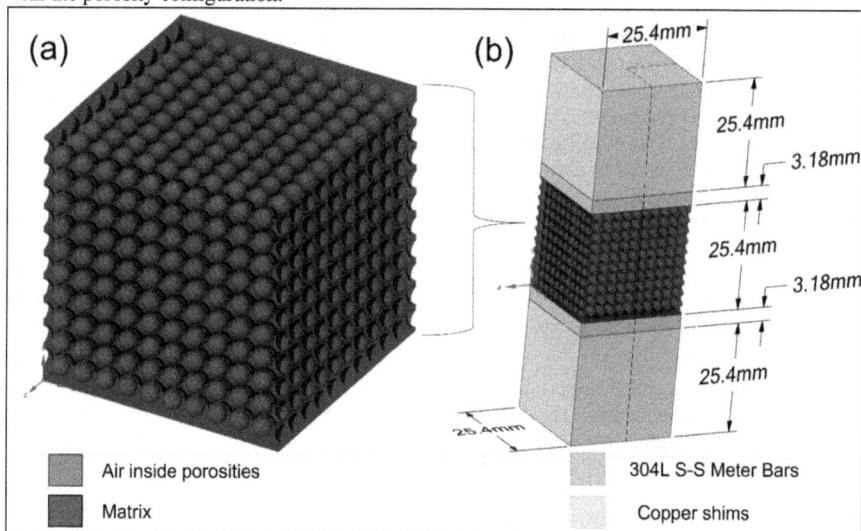

Fig. 1 S-S CMF model for thermal conductivity measurements using the GCH technique.

Fig. 1(b) depicts the computational model based on the GCH approach. Meter bars (304 L stainless steel), copper shims (copper), matrix and hollow spheres (316 L stainless steel), and air (within matrix and porosities) are the model's key constituents. The geometrical parameters shown in Fig. 1(b) are taken from the earlier work [4]. These components' thermal properties for the model can be found in Tables 1,2 and 3. Due to symmetry and unidirectional heat flow, only a quarter of the experimental setup was used as a representative volume for computational needs. Fig. 2 depicts the computational domain under consideration. The considered domain still represents bulk S-S CMF and is large enough to accommodate multiple spheres (5 across direction of heat flow and 11 in the direction of heat flow) in the body. The ANSYS Workbench's integrated meshing module was then used to mesh the derived geometric model. Within the problem domain, hexahedral elements were used to construct an organized grid.

Table 1. Thermal properties of the 316L and 304L steels at various temperature [6].

Temperature (°C)	316 L Stainless Steel			304 L Stainless Steel	
	Specific Heat (J kg^{-1} K^{-1})	Thermal Conductivity (W m^{-1} K^{-1})	Thermal Diffusivity (mm^2 s^{-1})	Specific Heat (J kg^{-1} K^{-1})	Thermal Conductivity (W m^{-1} K^{-1})
25	498.58	13.93	3.52	509.64	12.94
200	521.82	16.68	4.05	533.23	15.77
400	548.38	19.82	4.63	560.19	19.01
600	574.94	22.97	5.19	587.15	22.24
800	601.50	26.11	5.72	614.11	25.48
1000	628.06	29.25	6.23	641.08	28.72

Table 2. Thermal properties of copper at various temperature [7], [8].

Temperature (°C)	Specific Heat (J kg^{-1} K^{-1})	Thermal Conductivity (W m^{-1} K^{-1})
27	388.69	401
227	408.87	388
427	429.05	374
627	449.23	360
827	469.39	348
1027	489.57	338

Table 3. Thermal properties of air at various temperature [9].

Temperature (°C)	Density (kg m^{-3})	Specific Heat (J kg^{-1} K^{-1})	Thermal Conductivity (W m^{-1} K^{-1})	Viscosity (x 10^{-5} kg m^{-1} s^{-1})
25	1.184	1007	0.02551	1.849
200	0.7459	1023	0.03779	2.577
400	0.5243	1069	0.05015	3.261
600	0.4042	1115	0.06093	3.846
800	0.3289	1153	0.07037	4.362
1000	0.2772	1184	0.07868	4.826

Assumptions and Boundary Conditions. The meshed model was loaded into the FLUENT module of the ANSYS WORKBENCH. Here, the experiment's boundary conditions were applied, the necessary temperature gradients were measured, and the thermal conductivities at various temperatures were calculated. Applied boundary conditions are depicted schematically in Fig. 2, resembling the GCH method.

As mentioned earlier, the matrix was considered a bulk material in the computational domain. However, the PM method's matrix processes have porosities resembling a closed-cell foam structure. An analytical model for closed-cell metal foams was used to establish the matrix's thermophysical parameters and then entered into ANSYS FLUENT to represent the matrix's porous nature. In this work, the Maxwell-Eucken equation [10] was utilized to determine the thermal conductivity of the matrix used in the model:

$$k_m = \frac{k_s f_s + k_a f_a \frac{3k_s}{2k_s + k_a}}{f_s + f_a \frac{3k_s}{2k_s + k_a}} \qquad (1)$$

Here, k_m, k_s, and k_a are thermal conductivities (Wm^{-1} $^{\circ}$K^{-1}) of the solid bulk matrix, 316 L stainless steel, and air, respectively. Also, f_s and f_a are 316L stainless steel and air volume fractions within S-S CMF structure, respectively.

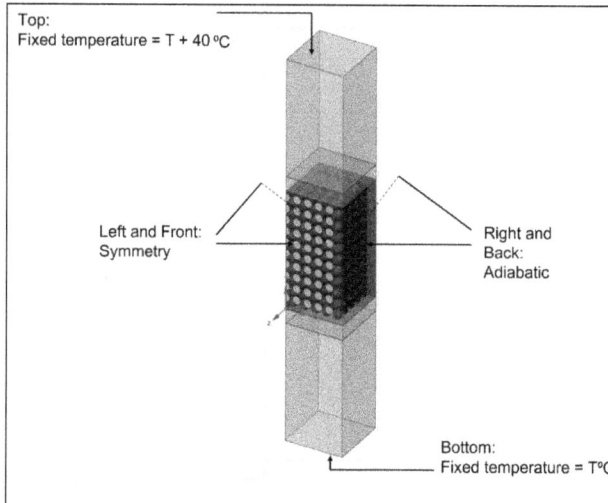

Fig. 2 Boundary conditions for the (2-mm sphere) S-S CMF as per the GCH experiments.

According to reports [11], metal foams have the same specific heat as their parent material; hence the specific heat of the matrix is equated to that of the solid 316L stainless steel.

Solution. ANSYS Fluent's steady-state condition solver was employed, and Fourier's Law was used to determine the thermal conductivity of the S-S CMF from 100 to 1000 $^{\circ}$C.

Results and discussions

Fig. 3 shows the temperature-dependent thermal conductivity of S-S CMF as per the results from the computational model along with the data from the GCH approach used in the earlier work [4]. The experimental data and the computational model both demonstrate that the S-S CMF exhibits significantly lower thermal conductivity than its main constituent, 316L stainless steel. Air trapped inside the sample's porosities, which has a substantially lower thermal conductivity than solid 316L stainless steel, is the cause of the lower thermal conductivity in CMF.

Since spheres make up 59 volume percent of S-S CMF, the matrix makes up 41 percent of the volume. With a wall thickness to sphere outer diameter ratio of 1/20, the air occupies about 73% of each sphere's cavity. As such, the air content inside the spheres is predicted to be about 43% of the total volume of the S-S CMF sample. In addition, from the 41% matrix that makes up the structure of CMF, around 46% of it is air in its micro-porosities, accounting for roughly 19% of air in the matrix (depending on the compaction of the matrix powder). Intrinsically, the total air in S-S CMF samples is about 19%+43%=62%. Air content in the matrix's microporosities (produced by powder metallurgy) and the air trapped inside the cavities of spheres lowers the S-S CMF thermal conductivity.

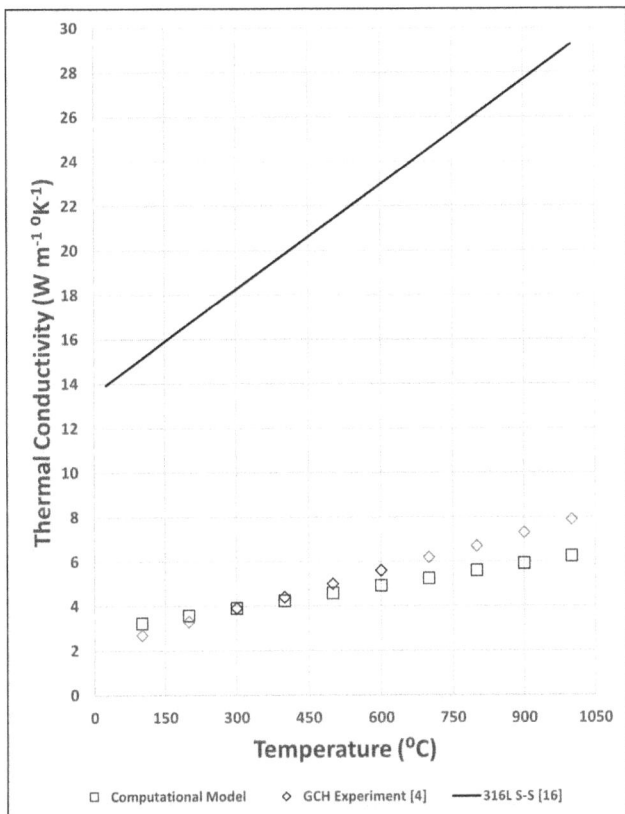

Fig. 3 Thermal conductivities of (2 mm sphere) S-S CMF by the computational model and the GCH experimental techniques. (Red data points are the extrapolated values from the GCH experiments).

Results also indicate differences in the thermal conductivity results between the experimental method used in the prior work [4] and the computational model—an average variance of 6% in the thermal conductivity readings between 300 and 600 °C. The extrapolated thermal conductivity data from the GCH experiment reveals an average departure of 15% from the computer modeling values for temperatures beyond 600 °C. Additionally, in contrast to the computational findings of the current work, we can see that the increase in thermal conductivity of S-S CMF with the temperature rise is more significant in the case of the GCH experimental technique [4]. Environmental factors, sample preparation, and measurement sensors may be to blame for the variations [12]. At high temperatures, convection, radiation, and non-unidirectional heat flow through the sample all contribute to increased measurement inaccuracy for the GCH technique [13]. For the GCH approach, extremely flat and parallel surfaces are necessary [12]. The air gaps between the sample surfaces and the hot and cold plates caused by roughness may significantly increase interfacial thermal resistance and a temperature drop. In the case of CMF, the contact surfaces are highly

Porous Metals and Metallic Foams: MetFoam 2023   Materials Research Forum LLC
Materials Research Proceedings 39 (2024) 17-24   https://doi.org/10.21741/9781644903094-3

porous (partial spheres and porous matrix); hence the air gaps can impact the thermal conductivity measurements.

The computational results were also compared with another experimental work [14] which employed the laser flash technique for finding the thermal conductivity of [(2 mm sphere) S-S CMF. It is well known that cellular materials' thermal conductivity is greatly influenced by their density [15]. Therefore, thermal conductivity results normalized by density are compared and shown in Fig. 4. The sample used for the computational studies was based on the GCH experimental work and hence was normalized by 2.7 g cm^{-3} [4], while the laser flash experiment samples had an average density of 2.6 g cm^{-3} [14]. Here we can see a close agreement between the computational results of this study with the experimental data of the laser flash experiments. Laser flash analysis is one of the most effective and adaptable ways to test thermal characteristics. Utilizing the appropriate sample holders and measurement settings determines thermal conductivity. The laser flash achieves High-accuracy temperature sensing using non-contact, non-destructive methods [16]. Hence it is devoid of the inaccuracies inherent in the GCH technique due to the high temperature and porous nature of CMF. Even though the computational model is based on the GCH technique, it avoids the environmental and sample contact surface effects by the applied ideal boundary conditions.

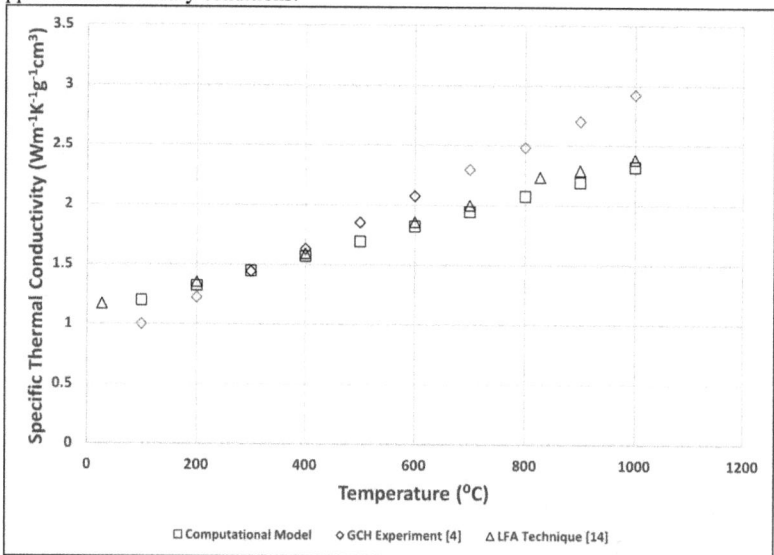

Fig. 4 Specific thermal conductivity of S-S CMF (Red data points are the extrapolated values from the GCH experiments).

We may therefore say that the computational model created here made fair assumptions and can successfully and reasonably predict the thermal conductivities of S-S CMF. The viability of S-S CMF for a specific thermal insulation application may be predicted very precisely using this model before it is produced. The model's sphere sizes and matrix porosity can be adjusted to estimate the thermal conductivity of different S-S CMF design iterations. This aids in choosing the best S-S CMF design for each application. As a result, the model can aid in reducing the cost of S-S CMF manufacture and experimental testing.

Conclusion

The following conclusions have been reached based on the findings of the experimental and modeling results:

- To forecast the thermal conductivity of S-S CMF, a straightforward physics-based computational model was created using ANSYS Fluent. The results of the computationally predicted thermal conductivity were compared with the earlier experiments and showed good agreement.
- It was shown that the S-S CMF's thermal conductivity is six times lower than that of its parent material, 316L S-S.
- The S-S CMF's porosities contain a high volume-fraction of air, which helps explain its low thermal conductivity.
- The S-S CMF thermal resistance can be altered by changing its porosity content in the matrix throughout adjusting the compaction pressure and by modifying the sphere-to-matrix ratios, and spheres geometry/characteristics.
- The cost of manufacturing and thermal experiments can be decreased by using the aforementioned computational model to help determine production parameters for S-S CMFs, such as the sphere sizes and porosity of the matrix.

Acknowledgment

This study is part of a project funded by the Department of Transportation (DOT) Pipeline and Hazardous Materials Safety Administration (PHMSA), project number PH957-20-0075.

References

[1] L. J. Vendra and A. Rabiei, "A study on aluminum–steel composite metal foam processed by casting," *Materials Science and Engineering: A*, vol. 465, no. 1–2, pp. 59–67, Sep. 2007. https://doi.org/10.1016/j.msea.2007.04.037

[2] B. P. Neville and A. Rabiei, "Composite metal foams processed through powder metallurgy," *Mater Des*, vol. 29, no. 2, 2008. https://doi.org/10.1016/j.matdes.2007.01.026

[3] W. H. Hsieh, J. Y. Wu, W. H. Shih, and W. C. Chiu, "Experimental investigation of heat-transfer characteristics of aluminum-foam heat sinks," *Int J Heat Mass Transf*, vol. 47, no. 23, pp. 5149–5157, Nov. 2004. https://doi.org/10.1016/j.ijheatmasstransfer.2004.04.037

[4] S. Chen, J. Marx, and A. Rabiei, "Experimental and computational studies on the thermal behavior and fire-retardant properties of composite metal foams," *International Journal of Thermal Sciences*, vol. 106, pp. 70–79, Aug. 2016. https://doi.org/10.1016/j.ijthermalsci.2016.03.005

[5] A. Rabiei and M. Garcia-Avila, "Effect of various parameters on properties of composite steel foams under variety of loading rates," *Materials Science and Engineering: A*, vol. 564, pp. 539–547, Mar. 2013. https://doi.org/10.1016/j.msea.2012.11.108

[6] C. S. Kim, "Thermophysical properties of stainless steels," 1975.

[7] B. Banerjee, An evaluation of plastic flow stress models for the simulation of high-temperature and high-strain-rate deformation of metals. 2005.

[8] J. G. Hust and A. B. Lankford, "Thermal conductivity of aluminum, copper, iron, and tungsten for temperatures from 1 K to the melting point," Jun. 1984.

[9] Y. A. Çengel, *Heat Transfer: A Practical Approach*. McGraw-Hill, 2003.

[10] A. Eucken, "Allgemeine Gesetzmäßigkeiten für das Wärmeleitvermögen verschiedener Stoffarten und Aggregatzustände," *Forschung auf dem Gebiete des Ingenieurwesens*, vol. 11, no. 1, pp. 6–20, Jan. 1940. https://doi.org/10.1007/BF02584103

[11] M. F. Ashby, A. Evans, N. A. Fleck, L. J. Gibson, J. W. Hutchinson, and H. N. G. Wadley, "Metal foams: a design guide," *Mater Des*, vol. 23, no. 1, p. 119, Feb. 2002. https://doi.org/10.1016/S0261-3069(01)00049-8

[12] "Standard Test Method for Thermal Conductivity of Solids by Means of the Guarded-Comparative-Longitudinal Heat Flow Technique 1". https://doi.org/10.1520/E1225-04.

[13] X.-H. An, J.-H. Cheng, H.-Q. Yin, L.-D. Xie, and P. Zhang, "Thermal conductivity of high temperature fluoride molten salt determined by laser flash technique," *Int J Heat Mass Transf*, vol. 90, pp. 872–877, Nov. 2015. https://doi.org/10.1016/j.ijheatmasstransfer.2015.07.042

[14] A. Rabiei, N. Amoafo-Yeboah, E. Huseboe, and C. Scemama, "A Study on Thermal Properties of Composite Metal Foams for Applications in Tank Cars Carrying Hazardous Materials," in *Minerals, Metals and Materials Series*, 2022. doi: 10.1007/978-3-030-92567-3_23

[15] A. N. Abramenko, A. S. Kalinichenko, Y. Burtser, V. A. Kalinichenko, S. A. Tanaeva, and I. P. Vasilenko, "Determination of the thermal conductivity of foam aluminum," *Journal of Engineering Physics and Thermophysics*, vol. 72, no. 3, pp. 369–373, May 1999. https://doi.org/10.1007/BF02699196

[16] S. Min, J. Blumm, and A. Lindemann, "A new laser flash system for measurement of the thermophysical properties," *Thermochim Acta*, vol. 455, no. 1–2, pp. 46–49, Apr. 2007. https://doi.org/10.1016/j.tca.2006.11.026

Porous Metals and Metallic Foams: MetFoam 2023
Materials Research Proceedings 39 (2024) 25-31

Materials Research Forum LLC
https://doi.org/10.21741/9781644903094-4

Custom Design to the Application of Open-Cellular Metal Structures

Claudia Drebenstedt[1,a*], Christian Hannemann[1,b], Jörg Hohlfeld[1,c],
Steve Siebeck[1,d], Thomas Hipke[1,e] and Dilay Kibaroglu[2,f]

[1]Fraunhofer IWU, Reichenhainer Str. 88, 09126 Chemnitz, Germany

[2]RWTH Aachen University, Intzestr. 1, 52072 Aachen, Germany

[a]claudia.drebenstedt@iwu.fraunhofer.de, [b]christian.hannemann@iwu.fraunhofer.de,
[c]joerg.hohlfeld@iwu.fraunhofer.de, [d]steve.siebeck@iwu.fraunhofer.de,
[e]thomas.hipke@iwu.fraunhofer.de, [f]dilay.kibaroglu@iehk.rwth-aachen.de

Keywords: Cellular Metals, Open Cellular Structures, Design, Investment Casting

Abstract. There are many potential applications for lightweight open-cellular metal structures, such as energy absorption, filtering, or thermal management. Such metallic open-cellular structures are often produced by additive manufacturing. Another option is the use of investment casting, for example by using open polymer foams as a template. By designing structures via Computer-Aided-Design (CAD), these can be used directly for additive manufacturing, either directly in metal or in wax as a template for casting. In this way the structure can be adapted very well to the needs, possible applications are shown in [1]. Using polymeric templates reduces the adaptability of the structure immensely. To use the full potential, it is necessary to develop the structure according to its future purpose. The 'ProZell' project is developing the basis for realizing such structures in high-manganese steels by investment casting using the lost wax process.

Introduction

The project ProZell aims to develop a programmable cellular metal structure made of high manganese steel. An open cellular structure is to be designed, consisting of a single unit cell, which can be scaled in size and number to fit a particular geometry depending on the later application. The aim is to combine a high lightweight potential with a high energy absorption potential. Two production routes and the influence on the resulting structure are to be compared, for example by means of compression tests. RWTH Aachen University is responsible for the additive manufacturing process and Fraunhofer IWU for the casting process.

Design and Process Setup

Based on the requirements described above, various designs for the cell geometry and the resulting structures were implemented in Catia. Some of them cannot be considered with regard to the feasibility of horizontal struts by additive manufacturing. Based on previous studies [2, 3] the structure shown on the left side of Fig. 1 was chosen as a starting point for a similar structure. The structure has vertical z-struts, which bend in a random direction when loaded in the z-direction.

Fig. 1: "f2cc,z structure" [2, 3] (left), CAD design of the ProZell structure (with base plate)

Porous Metals and Metallic Foams: MetFoam 2023   Materials Research Forum LLC
Materials Research Proceedings 39 (2024) 25-31   https://doi.org/10.21741/9781644903094-4

A structure with curved struts in the z-direction was designed for more predictable and controllable behavior (Fig. 1, right).

Process – Investment Casting

Wax printing with a ProJet 2500 (3D SYSTEMS) was used to create the positive placeholder for the structure. After removing the support wax in a heated isopropanol bath, the feeder was added. The structure was attached to a base plate via the feeder and placed in a cuvette. The next step was to fill the flask with the mold material using a KWS Investment Mixing Machine. After drying, the mold was placed in an oven to melt out the wax. The mold was bruned according to the pre-defined program and then held at the temperature selected for casting until it was transferred to the casting machine. A vacuum pressure casting machine, the BluePower INDUTHERM VC 480 V, was used for melting the material and casting. The chamber was flushed with argon to prevent oxidation. After cooling, the flask and the mold material were removed mechanically and additional with a high pressure water jet.

Materials

X30Mn22 was used for the casting as well as for the additive route. ProHT Steel and ProHT Platinum (Goodwin Refractory Services Limited) were used as mold material, the composition is stated in Table 1.

Table 1: Composition of ProHT mold material [4]

	Cristobalite	Quartz	Mono Ammonium Phosphate	Magnesia (MgO)	Fiber Glass
%	10 - 30	70 - 90	< 10	> 4	> 2

Furthermore, the mold materials Carath 1800 NC sf, Carath 1804 ULC (RATH Sales GmbH & Co. KG), and Keroyxd (Oxyd-Keramik GmbH & Co. KG) are used. The compositions are stated in Table 2.

Table 2: Composition of Carath 1800 NC sf, Carath 1804 ULC and Keroyxd [5, 6]

		Al_2O_3	SiO_2	CaO	MgO	Fe_2O_3	Na_2O
Carath 1800 NC sf	%	92.5	6.8	0.1	-	0.15	-
Carath 1804 ULC	%	97.8	0.15	1.5	-	0.15	-
Keroxyd	%	96.5	< 0.2	2.7	0.1	< 0.1	0.2

A zirconium-based coating from the manufacturer Ransom & Randolph was used as an additional intermediate layer. The wax model was dipped into a zirconium oxide slurry to a thickness of approximately 0.5 mm. The model was then sprinkled with zirconia sand and dried for at least 2 hours.

Experimental Setup

Initial trials showed that the cavity could not be completely filled, and strong reactions between the molding material and the molten X30Mn22 occurred. This resulted in very difficult / unfeasible demolding and geometric inaccuracy (Fig. 2). It is assumed that the molding reactions are based on the reaction of the manganese in the trip steel with the silicon in the molding material at higher temperatures and that compounds such as $MnSiO_3$ are formed. This has yet to proven by appropriate investigations.

Fig. 2: Results of the first trials for casting structures with high manganese steel

Due to the poor results shown in Fig. 2, it was decided to start with a geometry that was easier to evaluate: a wedge with three steps of different thicknesses (0.3 mm, 0.5 mm, 1 mm), each 15 mm long, in order to study the influence of the different parameters. The initial aim was to compare different temperatures and molding materials and the limits of feasibility in terms of melt flow in relation to thickness, which would allow an estimate of the possible strut diameters of the structure. The geometry specifications are shown in Fig. 3 and Fig. 4.

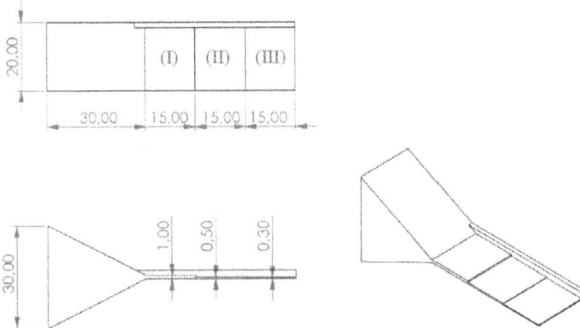

Fig. 3: Geometry of the wedge [1]

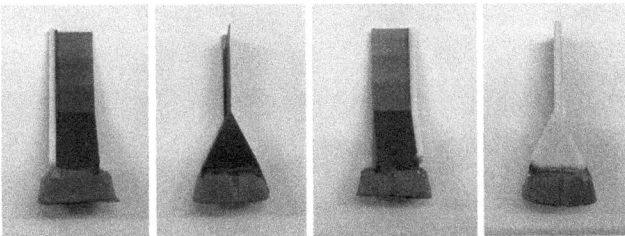

Fig. 4: Printed wax wedge with added feeder

Experimental

The casting molds were prepared as described (Process – Investment Casting). In addition to the high manganese steel, 316L steel was used as a reference as this material appeared to be easier to cast. The flow length was evaluated according to the steps 15 / 30 / 45 mm. The further steps were decided by evaluating the results of the previous steps.

Porous Metals and Metallic Foams: MetFoam 2023
Materials Research Proceedings 39 (2024) 25-31

Materials Research Forum LLC
https://doi.org/10.21741/9781644903094-4

Stage 1:

A DoE (design of experiments) was prepared for the first stage of castings. Two mold temperatures in the combination of with two ProHT mold materials and the 316L and high manganese steel were included as shown in Fig. 5.

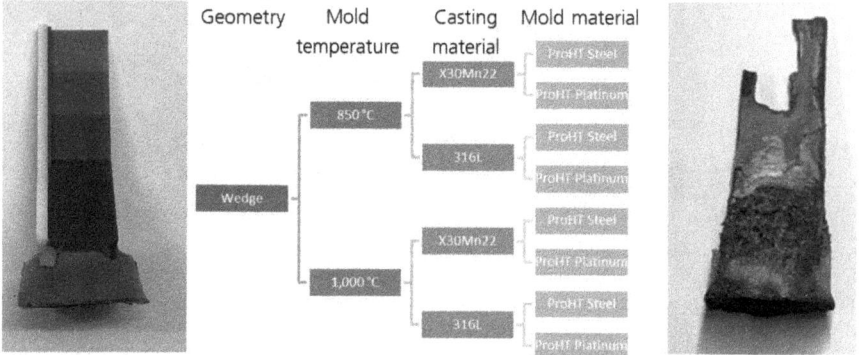

Fig. 5: Overview experiments stage 1

Results Stage 1:

The evaluation and the results of the experiments show a significant influence of the mold material on the flow length and thus on the degree of filling (0.5 mm step filled for the ProHT Steel mold, 0.3 step more than half filled), but no influence on the unwanted reactions between the mold and the casted steel (Fig. 6). The reactions between the high manganese steel and the mold material also caused a blowhole almost the size of the feeder itself (Fig. 8, center green frame), whereas for the casted 316L samples the surface, including the feeder, is smooth and straight (Fig. 8, left). Changing the mold temperature between 850 °C and 1,000 °C had no significant effect on any of the criteria evaluated.

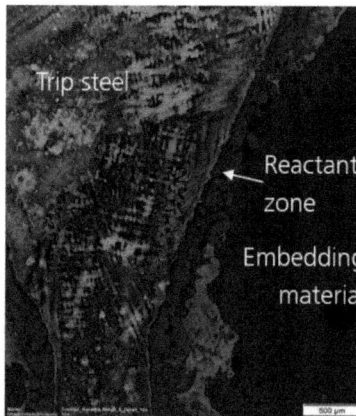

Fig. 6: Microsection of a Trip wedge with the reaction products at the surface

Stage 2:

In order to prevent the reactions described above, further trials were carried out with an additional intermediate layer on the wax models by adding a zirconium oxide slurry, Fig. 7. In addition, the mold temperature was reduced to 700 °C to investigate whether the reactions were reduced and whether there was an effect on the filling level. The mold material ProHT Steel was selected due to the better filling degree in stage 1.

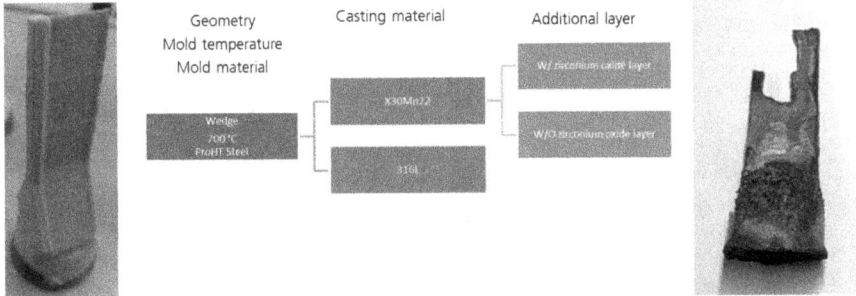

Fig. 7: Overview experiments stage 2

Results of Stage 2:
The degree of filling in 316L was slightly reduced due to the lower mold temperature of 700 °C. For the X30Mn22, there was little difference. The additional layer had no effect on the fill level. It reduces the reactions significantly, but the samples show that the layers had been damaged in some areas, so there are areas with a smooth surface and rough areas with reaction marks. The blowholes in the feeder are also smaller compared to the samples without the layer, as marked with a green frame in Fig. 8.

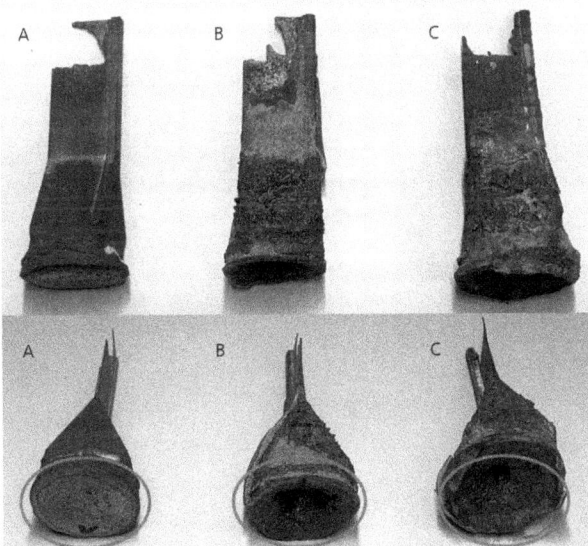

Fig. 8: (A) 316L, (B) X30Mn22, (C) X30Mn22 made with additional layer

Materials Research Forum LLC
https://doi.org/10.21741/9781644903094-4

Stage 3:
In order to evaluate and avoid the cracks in the interlayer, single and less vulnerable double layer samples were prepared and tested with mold temperatures of 850 °C and 1,000 °C, shown in Fig. 9. Only the X30Mn22 in combination with the ProHT steel was investigated at this stage.

Results Stage 3:
A higher degree of filling can be observed at 1,000 °C, but this could be due to a crack in the mold in two or three cuvettes. In general, the surface quality of the samples with two layers of zirconia slurry is much better and sharp edges are reproduced. Where the layer was intact there were no discernible reactions.

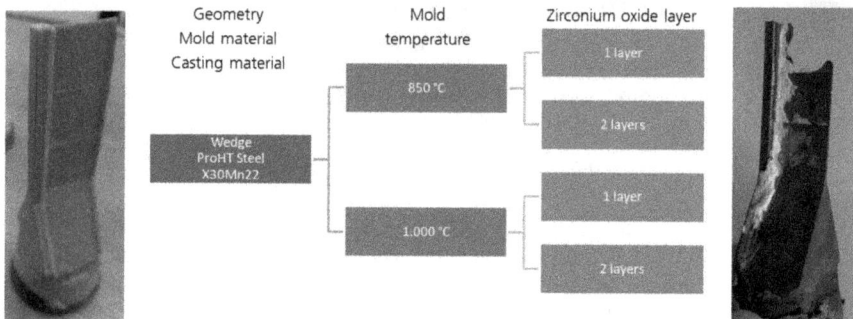

Fig. 9: Overview experiments stage 3, left: wax wedge with zirconium layer, right: sample made with 2 layers

Stage 4:
In order to avoid the reactions between the molding material and the trip steel, probably due to the high silicon content of the ProHT molding materials, 3 low silicon molding materials were tested in parallel with the zirconium layers, compare Fig. 10. Additional metallographic studies are ongoing.

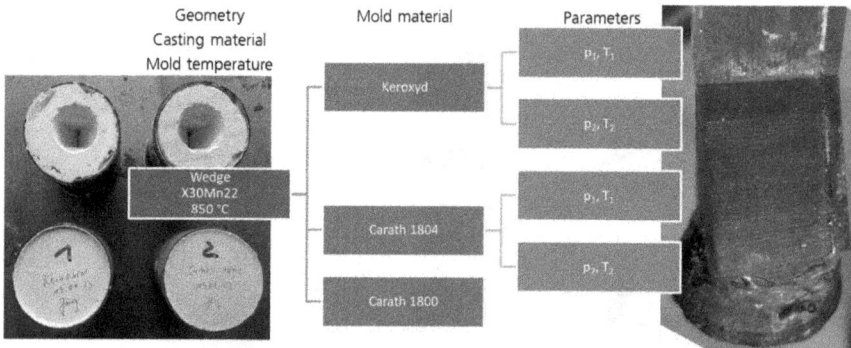

Fig. 10: Overview experiments stage 4, left: prepared molds, right: in Keroxyd casted Trip wedge

Results Stage 4:
Molds could not be manufactured using Carath 1800 NC sf ULC. However, Keroxyd and Carath 1804 ULC were successful in mold manufacturing with no noticeable melting reactions. The

surface quality of the molds was excellent, and even the horizontal lines from the wax's additive built-up were visible (Fig. 9, right). Keroxyd had a better filling degree / flow length (15 - 30 mm) compared to Carath 1804 ULC (0 - 15 mm). The material used for the mold was very sturdy, making it challenging to remove the part without causing damage, especially for delicate structures.

Summery and outlook

Casting experiments were conducted in four stages using various materials and parameters. The main criteria evaluated were flow length and reactions to the mold material. Some parameters were identified as significant. Treating the ProHT molds with an additional zirconia-based layer looks very promising. Additionally, using mold materials such as Keroxyd shows high potential. In the following steps, the mold material will be modified so that the mold can be removed more easily. 3-point flexure tests will be used to verify the effectiveness of this modification and transfer the results to the casting of the structures.

Acknowledgements

This work has been funded by Deutsche Forschungsgemeinschaft (DFG, German Research Foundation) within the project 'ProZell' (project ID 437986279).

We would like to thank NRU GmbH for supporting us by adding the zirconia layer to our wax structures.

References

[1] Hannemann, C., Uhlig, M., Hipke, T., & Meier, I. (2020). The Challenge of Open Cellular Metal Foam Production. In Proceedings of the 11th International Conference on Porous Metals and Metallic Foams (MetFoam 2019) (pp. 149-157). Springer International Publishing. https://doi.org/10.1007/978-3-030-42798-6_14

[2] Köhnen, P., Haase, C., Bültmann, J., Ziegler, S., Schleifenbaum, J. H., & Bleck, W. (2018). Mechanical properties and deformation behavior of additively manufactured lattice structures of stainless steel. Materials & Design, 145, 205-217. https://doi.org/10.1016/j.matdes.2018.02.062

[3] Köhnen, P., Létang, M., Voshage, M., Schleifenbaum, J. H., & Haase, C. (2019). Understanding the process-microstructure correlations for tailoring the mechanical properties of L-PBF produced austenitic advanced high strength steel. Additive Manufacturing, 30, 100914. https://doi.org/10.1016/j.addma.2019.100914

[4] PRO HT- Rev 07.pdf (srs-ltd.co.uk)

[5] https://www.rath-group.com/produkte/ungeformte-produkte/carath-d-carath-mc-dichte-betone, accessed: 07/11/2023

[6] https://www.oxyd-keramik.de/, accessed: 11/07/2023

Porous Metals and Metallic Foams: MetFoam 2023　　　　　　　　　　Materials Research Forum LLC
Materials Research Proceedings 39 (2024) 32-41　　　　　　　　　https://doi.org/10.21741/9781644903094-5

Analytical Determination of the Geometrical Properties of Open-Celled Metal Foams Under Compression

Esmari Maré[1,a*] and Sonia Woudberg[1,b]

[1]Applied Mathematics Division, Department of Mathematical Sciences, Stellenbosch University, Private Bag X1, Matieland, Stellenbosch, 7602, Western Cape, South Africa

[a]esmari.mare@gmail.com, [b]fidder@sun.ac.za

Keywords: Metal Foams, Permeability, Specific Surface Area, Compression, Darcy, Forchheimer

Abstract. Several studies in the literature have been devoted to the permeability prediction of metal foams, including that involving the Representative Unit Cell (RUC) model. The RUC modelling approach is an attractive modelling method due to the simple rectangular geometry, as well as its satisfying performance in comparison to other models and experimental data as proven elsewhere in the literature for porous media. The subject of compression of metal foams has been addressed elsewhere in the literature, but this study is the first to involve an all-inclusive analytical model where both streamwise and transverse compression are accounted for. The Darcy and Forchheimer permeability coefficients of the compressed foam (or three-strut) RUC model are presented. Furthermore, a geometric approach requiring measured geometric parameters and a combined geometric-kinetic approach involving measured permeability coefficients are included for determining the specific surface area. Geometric parameters required to determine the permeability and specific surface area predictions using the compressed foam RUC model include the uncompressed porosity, pore dimension and strut diameter, as well as the compression factor. The model is evaluated through comparison with available experimental data and empirical models obtained from the literature for compressed metal foams. The compressed RUC model predictions produce expected tendencies of geometrical parameters of metal foams under compression and the comparison with experimental data reveal satisfactory results.

Introduction

In the literature the topic of compression of foamlike media is often addressed. [1] and [2], for example, mentioned the advantages of compressed metal foams in that with compression, the density of the foam increases and consequently improves heat transfer and structural rigidity. The only geometric model that accommodates the structural transformation induced by compression as found by the author in the literature, is the streamwise compression transformation of the three-strut Representative Unit Cell (RUC) model [3,4]. Compression in the transverse direction (i.e. perpendicular to the direction of flow) is another consideration since it does appear in experimental studies available in the literature. For example, [5] investigated the controlling of microfluidic flow in microphysiological systems by compressing foams. Other examples of studies regarding the compression of metal foams include [1], [6], [7] and [8]. In this study the foam (or three-strut) RUC model is adjusted to accommodate the accompanied structural transformations. The model presented retains the rectangular nature, but is adjusted to facilitate different pore dimensions for each of the three principal directions. The model can consequently be applied to streamwise or transverse compression by implementing structural assumptions to the adjusted RUC model presented. From the adjusted model, expressions for the Darcy permeability, Forchheimer permeability and specific surface area can be determined. The equations for the specific surface area obtained using a geometric approach (i.e. in terms of pore-scale dimensions) is included as well as an example of a combined geometric-kinetic approach where the measured permeability

Porous Metals and Metallic Foams: MetFoam 2023   Materials Research Forum LLC
Materials Research Proceedings 39 (2024) 32-41   https://doi.org/10.21741/9781644903094-5

coefficient is required, rather than one of the pore-scale dimensions, in order to determine the specific surface area. Finally, an example of how the structurally transformed RUC model is applied to a specific case study of compression is illustrated using available experimental data for the permeability coefficients and specific surface area.

Model parameters

In this study, permeability coefficients are defined such that they are related to the pressure drop, as described by the Darcy-Forchheimer equation, in the following manner:

$$\frac{\Delta p}{L} = \frac{\mu}{K} q + \frac{\rho}{K_F} q^2 ,$$

(1)

where K and K_F are the permeability coefficients of the Darcy and Forchheimer regimes, respectively, $\Delta p/L$ is the pressure gradient, μ is the dynamic viscosity, ρ is the density of the fluid, and q is the superficial velocity. The compression factor, denoted by e_x, is introduced to relate the dimensions of the uncompressed models to the dimensions of the compressed models, where $0 < e_x \le 1$. When $e_x = 1$, the pore dimension corresponds to the uncompressed state. The compression factor is given by the ratio $e_x = h_f/h_o$, where h_o and h_f respectively denote the uncompressed and post-compression thickness of the porous medium sample in question [9]. The compression factor in this study is consequently determined as follows:

$$e_x = \frac{d_x}{d_{x_o}} ,$$

(2)

where d denotes the cell diameter of the RUC model and x respectively represents the subscripts '$\parallel$', '$\perp_1$' and '$\perp_2$' to denote compression in the streamwise direction and two possible transverse directions (e.g. $e_\parallel = d_\parallel/d_{\parallel_o}$ denotes the streamwise compression factor). The subscript 'o' denotes the uncompressed state. A general equation relating the porosity, ϵ, and compression factor was presented in [9,10], i.e.

$$\epsilon = 1 - \frac{1-\epsilon_o}{e_x} .$$

(3)

Eq. 3 is based on the assumption that the base area of the porous medium and the solid volume of the medium remains constant, as explained in [9]. In order to determine the values of the dimensions that are necessary to calculate the required predictions for the permeability coefficients and specific surface area provided by the adjusted RUC models, information regarding the direction of compression and compression factor are essential.

Adjusted three-strut RUC model

The three-strut RUC model with adjustable cell dimensions, is shown in Fig. 1. An expression for the porosity is determined first from which relations between the model dimensions can be obtained.

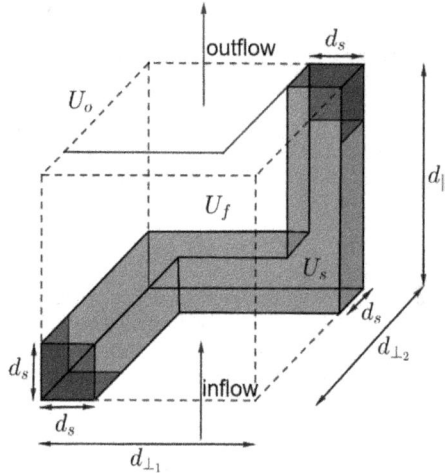

Fig. 1 Three-strut RUC model with adjustable dimensions

Model dimension relations. The total volume of the adjusted RUC model is $U_o = d_\parallel d_{\perp_1} d_{\perp_2}$ and the solid volume of the adjusted three-strut RUC model can be acquired from Fig. 1, which leads to the determination of the expression for the porosity, i.e.

$$\epsilon = 1 - \frac{U_s}{U_o} = 1 - \frac{d_s^2}{d_\parallel d_{\perp_1}} - \frac{d_s^2}{d_\parallel d_{\perp_2}} - \frac{d_s^2}{d_{\perp_1} d_{\perp_2}} + \frac{2\,d_s^3}{d_\parallel d_{\perp_1} d_{\perp_2}}. \tag{4}$$

Using Eq. 4, relations between the pore-scale dimensions and porosity can be attained.

Permeability prediction. In the studies of [3] and [10], an expression for the pressure drop of the Darcy flow regime is provided, as well as a derivation of the Darcy permeability coefficient of the RUC model. A similar pressure drop expression is provided in this study, with the difference that it is adapted to accommodate the adjustable dimensions of the adjusted three-strut RUC model, i.e.

$$\Delta p = \frac{S_{\parallel_1}\tau_{w_{\parallel_1}} + S_{\parallel_2}\tau_{w_{\parallel_2}} + S_{\perp_{1,1}}\tau_{w_{\perp_{1,1}}} + S_{\perp_{1,2}}\tau_{w_{\perp_{1,2}}} + S_{\perp_{2,1}}\tau_{w_{\perp_{2,1}}} + S_{\perp_{2,2}}\tau_{w_{\perp_{2,2}}}}{A_{p_\parallel}}. \tag{5}$$

In Eq. 5, $S_\parallel$, $S_{\perp_1}$ and $S_{\perp_2}$ denote the surface areas oriented in the directions relative to flow (which includes two transverse directions) and $\tau_\parallel$, $\tau_{\perp_1}$ and $\tau_{\perp_2}$ denote the magnitude of the corresponding average wall shear stresses. The subscripts of '1' and '2' in Eq. 5 (the second numerical subscript in the case of the transverse parameters) are utilized in order to distinguish between the two surface areas bounding the same channel as well as the magnitude of the corresponding average wall shear stresses. Using Fig. 1, the surface areas and average wall shear stress expressions are consequently determined to be

$$S_{\parallel_1} = 2d_s\big(d_{\perp_1} - d_s\big), \tag{6}$$
$$S_{\parallel_2} = 2d_s\big(d_{\perp_2} - d_s\big), \tag{7}$$

$$S_{\perp_{1,1}} = 2d_s\left(d_{\perp_2} - d_s\right), \tag{8}$$
$$S_{\perp_{1,2}} = 2d_s(d_\parallel - d_s), \tag{9}$$
$$S_{\perp_{2,1}} = 2d_s\left(d_{\perp_1} - d_s\right), \tag{10}$$
$$S_{\perp_{2,2}} = 2d_s(d_\parallel - d_s), \tag{11}$$

and

$$\tau_{w_{\parallel_1}} = \frac{6\mu w_\parallel}{d_{\perp_2} - d_s} = \frac{6d_{\perp_1}d_{\perp_2}}{(d_{\perp_1} - d_s)(d_{\perp_2} - d_s)^2}\,\mu q\,, \tag{12}$$

$$\tau_{w_{\parallel_2}} = \frac{6\mu w_\parallel}{d_{\perp_1} - d_s} = \frac{6d_{\perp_1}d_{\perp_2}}{(d_{\perp_1} - d_s)^2(d_{\perp_2} - d_s)}\,\mu q\,, \tag{13}$$

$$\tau_{w_{\perp_{1,1}}} = \frac{6\mu w_{\perp_1}}{d_\parallel - d_s} = \frac{6d_{\perp_1}d_{\perp_2}}{(d_\parallel - d_s)^2(d_{\perp_2} - d_s)}\,\mu q\,, \tag{14}$$

$$\tau_{w_{\perp_{1,2}}} = \frac{6\mu w_{\perp_1}}{d_{\perp_2} - d_s} = \frac{6d_{\perp_1}d_{\perp_2}}{(d_\parallel - d_s)(d_{\perp_2} - d_s)^2}\,\mu q\,, \tag{15}$$

$$\tau_{w_{\perp_{2,1}}} = \frac{6\mu w_{\perp_2}}{d_\parallel - d_s} = \frac{6d_{\perp_1}d_{\perp_2}}{(d_\parallel - d_s)^2(d_{\perp_1} - d_s)}\,\mu q\,, \tag{16}$$

$$\tau_{w_{\perp_{2,2}}} = \frac{6\mu w_{\perp_2}}{d_{\perp_1} - d_s} = \frac{6d_{\perp_1}d_{\perp_2}}{(d_\parallel - d_s)(d_{\perp_1} - d_s)^2}\,\mu q\,, \tag{17}$$

respectively. $w_\parallel$, $w_{\perp_1}$ and $w_{\perp_2}$ in Eqs. 12 to 17 denote the average channel velocities in the indicated directions. The expression for the streamwise cross-sectional flow area ($A_{p_\parallel}$) is furthermore given by

$$A_{p_\parallel} = \left(d_{\perp_1} - d_s\right)\left(d_{\perp_2} - d_s\right). \tag{18}$$

Substituting the equations for the surface areas as given by Eqs. 6 to 11, Eqs. 12 to 17 and Eq. 18 into Eq. 5, leads to the following equation for the pressure gradient of the Darcy flow regime:

$$\frac{\Delta p}{d_\parallel} = \frac{24\,d_s\,d_{\perp_1}d_{\perp_2}\mu q}{d_\parallel(d_{\perp_1} - d_s)(d_{\perp_2} - d_s)}\left[\frac{1}{(d_\parallel - d_s)^2} + \frac{1}{(d_{\perp_1} - d_s)^2} + \frac{1}{(d_{\perp_2} - d_s)^2}\right]. \tag{19}$$

The equation used to determine the pressure gradient of the Forchheimer flow regime is the same as the equation utilized for this purpose by [11], i.e.

$$-\nabla\langle p\rangle_f = \frac{S_{\text{face}}}{\epsilon U_o}\rho w_\parallel{}^2\hat{\underline{n}} = \frac{d_{\perp_1}d_{\perp_2}(d_s d_{\perp_1} - d_s{}^2)}{\epsilon d_\parallel(d_{\perp_1} - d_s)^2(d_{\perp_2} - d_s)^2}\,\rho q^2\hat{\underline{n}}\,, \tag{20}$$

where S_{face} is the cross-sectional solid area that faces upstream, as determined from Fig. 1. The Ergun-type equation consequently leads to

$$\frac{\Delta p}{d_\parallel} = \frac{24\,d_s\,d_{\perp_1}d_{\perp_2}}{d_\parallel(d_{\perp_1} - d_s)(d_{\perp_2} - d_s)}\left[\frac{1}{(d_\parallel - d_s)^2} + \frac{1}{(d_{\perp_1} - d_s)^2} + \frac{1}{(d_{\perp_2} - d_s)^2}\right]\mu q +$$
$$\frac{d_{\perp_1}d_{\perp_2}(d_s d_{\perp_1} - d_s{}^2)}{\epsilon d_\parallel(d_{\perp_1} - d_s)^2(d_{\perp_2} - d_s)^2}\,\rho q^2\,, \tag{21}$$

and comparing Eq. 21 with Eq. 1 yields the expressions of the Darcy and Forchheimer permeability coefficients of the adjusted three-strut RUC model, respectively denoted by K and K_F, i.e.

Materials Research Forum LLC

https://doi.org/10.21741/9781644903094-5

$$K = \frac{d_{\parallel}(d_{\perp_1} - d_s)(d_{\perp_2} - d_s)}{24\, d_s\, d_{\perp_1} d_{\perp_2}} \left[\frac{1}{(d_{\parallel} - d_s)^2} + \frac{1}{(d_{\perp_1} - d_s)^2} + \frac{1}{(d_{\perp_2} - d_s)^2} \right]^{-1}, \tag{22}$$

and

$$K_F = \frac{\epsilon d_{\parallel}(d_{\perp_1} - d_s)^2 (d_{\perp_2} - d_s)^2}{d_{\perp_1} d_{\perp_2}(d_s d_{\perp_1} - d_s^2)}. \tag{23}$$

Specific surface area: geometric approach. The total surface area of the adjusted three-strut RUC model can be determined by adding all the individual surface areas given by Eqs. 6 to 11. Dividing the result by the total RUC volume, the equation for the specific surface area can be determined, i.e.

$$S_v = \frac{4\, d_s}{d_{\parallel} d_{\perp_1} d_{\perp_2}} \left[(d_{\parallel} - d_s) + (d_{\perp_1} - d_s) + (d_{\perp_2} - d_s) \right]. \tag{24}$$

Due to the dependence of the specific surface area on the geometry of the three-strut RUC model as shown in Fig. 1, the porosity equation (Eq. 4) relates the dimensions in Eq. 24.

Specific surface area: combined approach. In the combined kinetic-geometric approach for determining the specific surface area, also referred to as simply the combined approach, a permeability coefficient is utilized to determine S_v instead of one of the pore-scale dimensions. Either one of the pore-scale dimensions in Eq. 24 can be replaced. The substitution of an expression for K instead of the pore-scale dimension $d_{\parallel}$ is considered in this study as an example of acquiring the specific surface area using a combined approach. First, the expression for S_v (i.e. Eq. 24), needs to be rearranged to determine $d_{\parallel}$ in terms of the specific surface area. This yields

$$d_{\parallel} = \frac{4s - 4d_s^2}{d_{\perp_1} d_{\perp_2} S_v - 4d_s}, \tag{25}$$

where $s = d_s\left[(d_{\perp_1} - d_s) + (d_{\perp_2} - d_s) \right]$. Eq. 25 is then substituted into Eq. 19 and rearranged in order to obtain the following third degree polynomial in S_v from which S_v can be solved:

$$a' S_v^3 + b' S_v^2 + c' S_v + g' = 0,$$

where

$$
\begin{aligned}
a' &= -K d_{\perp_1}^3 d_{\perp_2}^3 (1 + n d_s^2) \\
b' &= 12 K d_s\, d_{\perp_1}^2 d_{\perp_2}^2 (1 + n d_s^2) + 4 d_s d_{\perp_1}^2 d_{\perp_2}^2 (m d_s + 2nK)(s - d_s^2) \\
c' &= -48 K d_s^2 d_{\perp_1} d_{\perp_2} (1 + n d_s^2) - 32 d_s^2 d_{\perp_1} d_{\perp_2} (m d_s + 2nK)(s - d_s^2) \\
&\quad - 16 d_{\perp_1} d_{\perp_2} (2 m d_s + nK)(s - d_s^2)^2 \\
g' &= 64 K d_s^3 (1 + n d_s^2) + 64 d_s^3 (m d_s + 2nK)(s - d_s^2) + 64 d_s (2 m d_s + nK)(s - d_s^2)^2 \\
&\quad + 64 m (s - d_s^2)^3.
\end{aligned}
\tag{26}
$$

with $m = \dfrac{(d_{\perp_1} - d_s)(d_{\perp_2} - d_s)}{24\, d_s\, d_{\perp_1} d_{\perp_2}}$ and $n = \dfrac{1}{(d_{\perp_1} - d_s)^2} + \dfrac{1}{(d_{\perp_2} - d_s)^2}$. It should be noted that in order to determine the specific surface area by making use of the combined approach, experimental permeability data is required, e.g. experimental Darcy permeability data should be substituted into

Eq. 26 in order to determine the specific surface area of the adjusted three-strut RUC model when using Eq. 26.

Model application and validation

In this section, the manner in which the compressed three-strut RUC model can be utilized and validated, using available experimental data obtained from [12], is shown and the predictions acquired are compared to that of the corresponding uncompressed model and experimental data. [12] investigated the correlations of the pressure drop for air flow through different foam samples, of which some were compressed. Permeability and specific surface area data for four of these foam samples that were compressed in the streamwise direction are shown in Table 1.

Table 1 Experimental data obtained from [12] for aluminum foams

ϵ_o []	ϵ []	e []	d_{p_o} [μm]	$K \times 10^9$ [m^2]	$K_F \times 10^3$ [m]	S_v [m^{-1}]
0.921	0.679	0.246	1300	0.66	0.323	5104.3
0.921	0.774	0.350	1300	0.10	0.500	3593.7
0.922	0.682	0.245	2500	0.10	0.588	3169.3
0.922	0.794	0.379	2500	0.21	0.00012	2053.1

The uncompressed porosity and PPI numbers (and hence a way to obtain the uncompressed pore diameter) were provided by [12], along with corresponding porosity and compression factor data, as shown in Table 1. The compression factor data, however, was determined by [12] by using Eq. 3 which assumes no lateral expansion of the foam under compression, as mentioned in the model parameters section of this study. Consequently, this condition is assumed in this study. Due to the media samples being compressed in the streamwise direction, it is deduced from Eq. 2 that

$$d_\parallel = e_\parallel d_{\parallel_o} . \tag{27}$$

No value for the uncompressed strut diameter d_{s_o} or the uncompressed cell diameter $d_{\parallel_o}$ was provided by [12] from which all the parameters associated with the uncompressed state can be determined. Therefore, it is assumed that the uncompressed state of the three-strut RUC model is isotropic. This is similar to the assumption made by [13] in the implementation of the streamwise compressed three-strut RUC model in the application to data provided by [14]. Consequently, $d_{\parallel_o} = d_{\perp_{1_o}} = d_{\perp_{2_o}} = d_o$ and d_o can be obtained using the following relations between the dimensions of the isotropic three-strut RUC model:

$$d_o = \frac{2d_{p_o}}{3 - \psi_o} , \tag{28}$$

where $d_{s_o} = d_o - d_{p_o}$ and ψ_o denoted a geometric factor, given by [3]

$$\psi_o = 2 + 2\cos\left[\frac{4\pi}{3} + \frac{1}{3}\cos^{-1}(2\epsilon_o - 1)\right]. \tag{29}$$

It is furthermore assumed that $d_{\perp_1} = d_{\perp_2} = d_o$, due to no lateral expansion. Due to the decrease in porosity combined with the assumption of constant solid volume during compression, as well as the condition of no lateral expansion, d_s increases with compression. The value of d_s can be determined from the following rearrangement of Eq. 4:

$$2d_s{}^3 + (-2d_o - d_\parallel)d_s{}^2 + d_o{}^2 d_\parallel(1 - \epsilon) = 0 .\tag{30}$$

The Darcy permeability coefficient, Forchheimer permeability coefficient and specific surface area obtained using the geometric approach are thus, respectively, given by

$$K = \frac{d_\parallel(d_o - d_s)^2}{24 d_s d_o{}^2}\left[\frac{1}{(d_\parallel - d_s)^2} + \frac{2}{(d_o - d_s)^2}\right]^{-1},\tag{31}$$

$$K_F = \frac{\epsilon d_\parallel (d_o - d_s)^3}{2 d_s d_o{}^2},\tag{32}$$

and

$$S_v = \frac{4 d_s}{d_\parallel d_o{}^2}\left[(d_\parallel - d_s) + 2(d_o - d_s)\right].\tag{33}$$

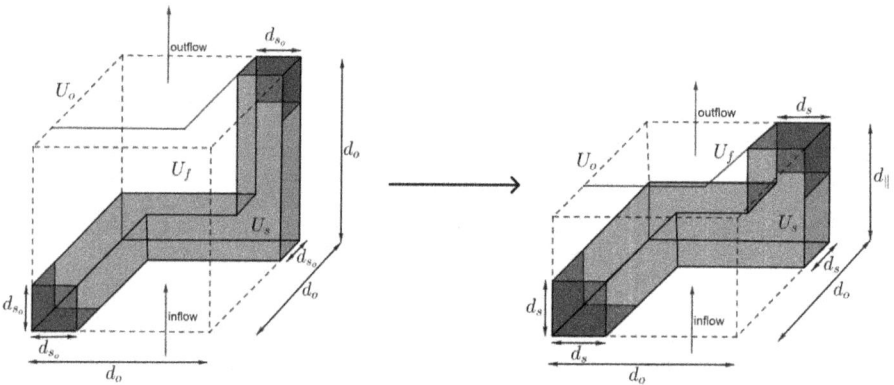

Fig. 2 Three-strut RUC model under streamwise compression without lateral expansion

The combined approach for determining the specific surface area can be utilized due to the availability of both permeability and specific surface area data by replacing $d_\parallel$ with K. A visual representation of the streamwise compressed three-strut RUC model subject to deformation (with no lateral expansion) is shown in Fig. 2. In Figs. 3 to 5 the predictions and experimental data of the Darcy permeability coefficient, Forchheimer permeability coefficient and specific surface area (using both the geometric and combined approach of attaining S_v), utilizing the data provided in Table 1 together with Eqs. 27 to 33, are shown, respectively. Empirical expressions for the Darcy permeability and form drag coefficients were furthermore provided by [2] and [12], from which the expression for the form drag coefficient can be utilized to obtain the prediction for the Forchheimer permeability coefficient. The permeability predictions acquired and corresponding data for the foam samples of [12] are included in Figs. 3 and 4. In Fig. 3 the compressed three-strut RUC model predictions provide closer correspondence to the experimental data than the uncompressed model predictions and In Fig. 4 it can be seen that the compressed RUC model corresponds excellently with the experimental data and performs better than the uncompressed and empirical models considered. The values provided by the experimental data are lower than that of the uncompressed model for both permeability coefficients, as expected, with the compressed three-strut RUC model predictions being in turn lower than that of the uncompressed model predictions. It is also noted that the compressed three-strut RUC model also provides closer predictions to the experimental data than that of the empirical models of [2] and [10], which lends favor to the adjusted foam RUC model presented in this study.

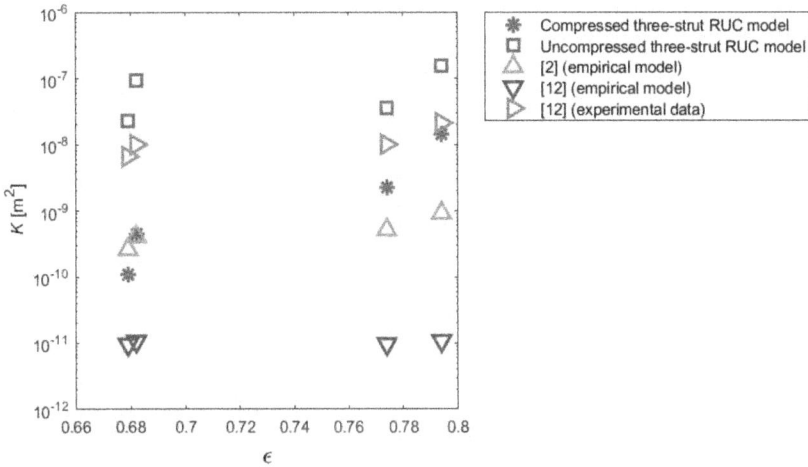

Fig. 3 Darcy permeability prediction versus porosity of RUC models and metal foam experimental data obtained from [10] under compression

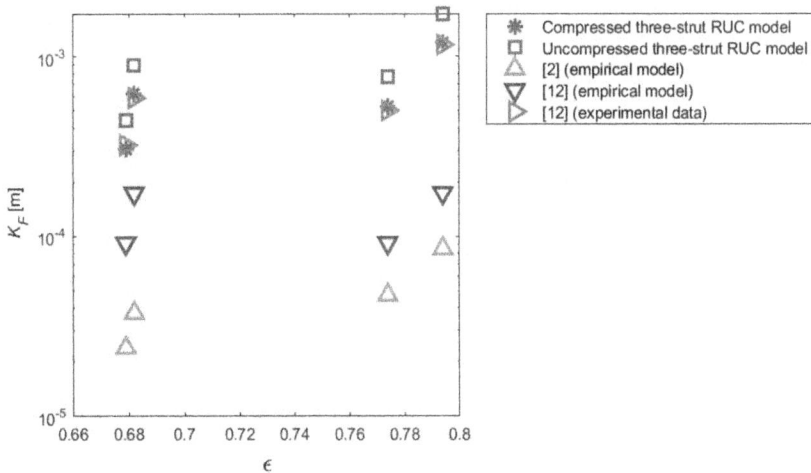

Fig. 4 Forchheimer permeability prediction versus porosity of RUC models and metal foam experimental data obtained from [10] under compression

Porous Metals and Metallic Foams: MetFoam 2023
Materials Research Proceedings 39 (2024) 32-41

Materials Research Forum LLC
https://doi.org/10.21741/9781644903094-5

In Fig. 5 the compressed three-strut RUC model once again corresponds closer to the experimental data, resulting from the predictions obtained using a geometric approach. The specific surface area predictions of the compressed RUC model acquired using a combined approach, however, corresponds closer to the predictions of the uncompressed model, which significantly under predicts the data. This may be due to the uncertainty already incorporated into the permeability prediction which is used as input to the combined approach. The combined approach can however be used to obtain S_v values of the correct order of magnitude, should specific surface area data be required but is unavailable.

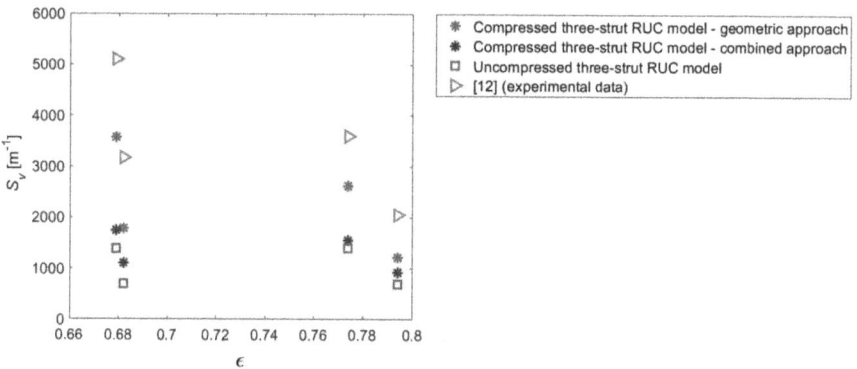

Fig. 5 Specific surface area prediction versus porosity of RUC models and metal foam experimental data obtained from [10] under compression

Summary

An adjusted foam (or three-strut) RUC model was presented that accommodates the parameter adjustments associated with compressed foamlike media and can be utilized in the application of both streamwise and transverse compression. Predictions for the Darcy and Forchheimer permeability coefficients, based on the adjusted three-strut RUC model, have been proposed, as well as the specific surface area prediction in terms of the linear pore-scale dimensions of the foam. A combined method for obtaining the specific surface area using the Darcy permeability, was furthermore included, where the experimentally obtained permeability can be utilized instead of one of the pore-scale dimensions, should specific surface area values be required and data is unavailable. Validation of the compressed RUC model was presented for streamwise compression with no lateral expansion and demonstrated using available experimental data. Comparison with experimental data for compressed foams revealed that the compressed model shows an improvement in predictions for the Darcy and Forchheimer coefficient values as well as specific surface area predictions when compared to that of the uncompressed model. Further model validation can be supported through the availability of additional experimental data that include measured characteristics of the uncompressed foam (i.e. the strut diameter of the uncompressed state) as well as permeability and specific surface area values of the compressed foam.

References

[1] B.V. Antohe, J.L. Lage, D.C. Price, R.M. Weber, Experimental determination of permeability and inertia coefficients of mechanically compressed aluminum porous matrices, ASME. J. Fluids Eng. Vol. 119(2) (1997) 404-412. https://doi.org/10.1115/1.2819148

[2] N. Dukhan, R. Picón-Feliciano, Á.R. Álvarez-Hernández, Air flow through compressed and uncompressed aluminum foam: measurements and correlations, ASME. J. Fluids Eng. 128(5) (2006) 1004-1012. https://doi.org/10.1115/1.2236132

[3] S. Woudberg, M.C. van Heyningen, L. Le Coq, J. Legrand, J.P. Du Plessis, Analytical determination of the effect of compression on the permeability of fibrous porous media, Chem. Eng. Sci. 112 (2014) 108-115. https://doi.org/10.1016/j.ces.2014.03.013

[4] S. Woudberg, E. Maré, M.C. van Heyningen, F. Theron, L. Le Coq, Predicting the permeability and specific surface area of compressed and uncompressed fibrous media including the Klinkenberg effect, Powder Technol. 377 (2021) 488-505. https://doi.org/10.1016/j.powtec.2020.08.081.

[5] S. Hong, Y. Jung, R. Yen, H.F. Chan, K.W. Leong, G.A. Truskey, X. Zhao, Magnetoactive sponges for dynamic control of microfluidic flow patters in microphysiological systems, Lab Chip 14(3) (2014) 514-521. http://dx.doi.org/10.1039/C3LC51076J

[6] K. Boomsma, D. Poulikakos, The effects of compression and pore size variations on the liquid flow characteristics in metal foams, ASME. J. Fluids Eng. 124(1) (2002) 263-272. https://doi.org/10.1115/1.1429637

[7] J. Hwang, G. Hwang, R. Yeh, C. Chao, Measurement of interstitial convective heat transfer and frictional drag for flow across metal foams, ASME. J. Heat Transfer 124(1) 120-129. http://doi.org/10.1115/1.1416690

[8] M.A. Dawson, J.T. Germaine, L.J. Gibson, Permeability of open-cell foams under compressive strain, Int J. Solids Struct. 44(16) (2007) 5133-5145. https://doi.org/10.1016/j.ijsolstr.2006.12.025

[9] M.C. van Heyningen, Investigating the effect of compression on the permeability of fibrous porous media, MSc Thesis, Stellenbosch University, South Africa, 2014.

[10] S. Woudberg, Permeability prediction of an analytical pore-scale model for layered and isotropic fibrous porous media, Chem. Eng. Sci. 164 (2017) 232-245. https://doi.org/10.1016/j.ces.2017.01.061

[11] S. Woudberg, J.P. Du Plessis, An analytical Ergun-type equation for porous foams, Chem. Eng. Sci. 148 (2016) 44-54. https://doi.org/10.1016/j.ces.2016.03.013

[12] N. Dukhan, Correlations for the pressure drop for flow through metal foam, Exp Fluids 41 (2006) 665-672. https://doi.org/10.1007/s00348-006-0194-x

[13] S. Woudberg, Investigation of the effect of compression on a soft fibrous porous medium, Computational Methods in Multiphase Flow VII 79, 2013, pp.351.

[14] H.D. Akaydin, A. Pierides, S. Weinbaum, Y. Andreopoulos, Permeability of soft porous media under one-dimensional compaction, Chem. Eng. Sci. 66(1) (2011) 1-14. https://doi.org/10.1016/j.ces.2010.09.017

Porous Metals and Metallic Foams: MetFoam 2023
Materials Research Proceedings 39 (2024) 42-50

Materials Research Forum LLC
https://doi.org/10.21741/9781644903094-6

Numerical Investigation on Deformation Behavior of Aluminium Foams with *in situ* Composite Particles

M. Rakesh[1,a], A. Tewari[1] and S. Karagadde[1,b]

[1]Department of Mechanical Engineering, Indian Institute of Technology Bombay, Mumbai, India

[a]merugurakesh94@gmail.com, [b]s.karagadde@iitb.ac.in

Keywords: Metal Matrix Composite, *in situ* Metal Foam, Finite Element Analysis, Liquid Metallurgy Process, Energy Absorption

Abstract. Metal foams are cellular solids with high stiffness, high strength and superior energy absorption capacity. In liquid metallurgy, foams are processed by foaming the molten metal with the addition of foaming agents and stabilized by the presence of particles which also strengthen the cell walls. An analysis of the deformation behaviour of foams in the presence of these stabilizing particles is essential to the mechanism of energy absorption. In the present study, the effect of the particle distribution on the deformation behaviour of the closed-cell aluminium foams was investigated using Finite Element Analysis (FEA) in Abaqus© software. The experimental data were used to model the distribution of particles in the matrix and the foam model. The simulation results were validated with the experimental results. The uniform distribution of particles in the matrix resulted in lower stress concentration and enhanced the mechanical performance of composite material and the metal foam.

Introduction

Metal foams are lightweight materials with high stiffness, high specific strength-to-weight ratio and superior specific energy absorption capacity [1–3]. Processing of metal foams through the liquid metallurgy route involves, melting and foaming of molten metal or alloy by the addition of foaming agents such as TiH_2 or $CaCO_3$ in the presence of stabilizing particles to obtain a uniform foam structure. The properties of foam mainly depend on the structure of the foam, distribution of pore size and cell wall thickness, and the type of particles used and their distribution for stabilization [4, 5]. The particles used for the foam stabilization were added externally (*ex situ*) or generated internally using *in situ* chemical reaction with the matrix material. It is well known that *in situ* particles exhibit superior bonding strength, wetting behaviour and nucleation potency with the primary phase over the *ex situ* particles and efficiently stabilise the foam structure [6–8]. Nevertheless, the particles tend to form clusters near the cell walls and plateau borders when the metal foams are processed through the liquid metallurgy route. Studying the role of these particle clusters on the properties of foam is necessary to understand the deformation behaviour of foam.

FEA is a numerical technique widely used to analyse the elastic-plastic behaviour of heterogeneous materials such as Metal Matrix Composites (MMC) and metal foams [9]. Ma *et al* numerically investigated the mechanical and fracture behaviour of particle reinforce A356 composite with 2D representative volume element (RVE) models and reported that, the simulation results closely matched the experimental results with the use of damage evolution criteria [10, 11]. Antunes *et al* developed a new model for estimating the flexural properties of epoxy and polyester resin-based syntactic foams and concluded that the flexural stiffness decreased with the filler volume fraction and epoxy-based foams outperformed polyester foams [12]. Nammi *et al* analysed the quasi-static loading of closed-cell aluminium foams using the repeating unit cell method based on the tetrakaidekahedral model and accurately estimated the crushing resistance of the foam [13]. Perez *et al* estimated the properties of CNT-reinforced aluminium metallic foams and found that Young's modulus prediction using FEA analysis was within the theoretical values and smaller

Porous Metals and Metallic Foams: MetFoam 2023   Materials Research Forum LLC
Materials Research Proceedings 39 (2024) 42-50   https://doi.org/10.21741/9781644903094-6

pores lead to lower values of elastic modulus [14]. Karan *et al* predicted the compressive deformation behaviour of LM 13 aluminium foam in accordance with the experimental results. However, the role of particle distribution on the properties of metal foams has not been studied in detail in the past which is essential to understand the deformation of behaviour of metal foam during mechanical testing in turn to optimise the process parameters.

In the present study, the effect of particle clusters on the deformation behaviour of metal foams was investigated using 2D FEM (Finite Element Method). 2D FEM was widely used to study the stress and strain distributions for multi-scale simulations, especially for MMCs [15], but it was limited to planar deformations with a strong dependency on the domain size. However, it will provide insights into the initial approximations of stress-strain behaviour of multi-scale models. Among various types of *in situ* particles that were analysed in the previous studies, *in situ* Al$_3$Zr particles that possess low density (4.11 g/cm^3), high stiffness (150 GPa) and high melting point (1580 °C), have shown better stabilisation of the foam with uniform pore distribution and good nucleation potency with the primary phase [16–18] and are employed in the present investigation. 2D RVE models were used to analyse the fracture behaviour of MMC with *in situ* Al$_3$Zr particles. The constitutive behaviour of MMCs was used to perform the deformation simulations of foam structures.

Numerical methodology
The numerical investigation was carried out in two stages. Initially, the properties of MMC with uniform and clustered distributions of particles were analysed. In the later stage, the constitute behaviour of MMCs obtained from the simulation was used to study the deformation behaviour of metal foams with compression simulation.

a). Simulation of MMC
Different sizes of 2D RVE models were employed to analyse the deformation and damage behaviour of A356 with *in situ* Al$_3$Zr particles. The 2D RVE models were generated using in-house Python code that was implemented using Abaqus© software. The particle generation takes place if the particle was not overlapping with the previously existing particle and in contact with the edges of the RVE. This operation was repeated till the desired volume fraction of the particle was obtained. The particle distribution varied by varying the coefficient of variation of the nearest neighbour distance (COV$_d$) of the particle which is the ratio of the standard deviation of the particle size distribution to the mean nearest neighbour distance of the particle [19].

The MMC samples were prepared for metallography to measure the size distribution of Al$_3$Zr particles in the matrix. From the metallography of the Al$_3$Zr particles, the shape of the particle was identified as a rectangle of length 11.2± 0.62 μm and width of 7.3±0.61 μm. For modelling cluster distribution, a COV$_d$ value of 0.325 was used, while for uniform distribution a value of 0.062 was employed. The properties of A356 alloy are adopted from the literature [20]. The properties of particles are calculated from the reverse solution algorithm proposed by Dao *et al* [21]. The input parameters (Nano-indentation results) required for the algorithm are shown in Table 1, where v is Poisson's ratio, C is the loading curvature, h$_r$/h$_m$ is the ratio of residual indentation depth to maximum indentation depth, P_{ave} is the average load and P_m is the maximum load. The parameters calculated from the reverse solution algorithms are shown in Table 2. Where $\sigma_{0.033}$ is actual stress at 0.033 strain, E is Young's modulus, n is strain hardening exponent and σ_y is yield stress. Fig. 1 shows the stress-strain behaviour of the matrix and particle implemented in the simulation. The periodic boundary conditions and loading on the RVE with uniform and clustered are shown in Fig. 2a and 2b respectively. For modelling the damage evolution in MMC, ductile damage criteria were assigned for matrix material while brittle cracking criteria were assigned for particle material.

Porous Metals and Metallic Foams: MetFoam 2023
Materials Research Forum LLC
Materials Research Proceedings 39 (2024) 42-50
https://doi.org/10.21741/9781644903094-6

The stress-strain behaviour was obtained from the homogenization method proposed by Okereke and Akpoyomare's work [22, 23].

Table 1. Input parameters for the Al₃Zr phase used for the reverse solution algorithm

v	C/GPa	h_r/h_m	P_{ave} (GPa)	P_m (mN)
0.185	198	0.757	7.204	10

Table 2. Parameters obtained for the Al₃Zr phase used for the reverse solution algorithm

$\sigma_{0.033}$ (Mpa)	E (GPa)	n	σ_y (MPa)
1864.77	146.17	0.651	283.5

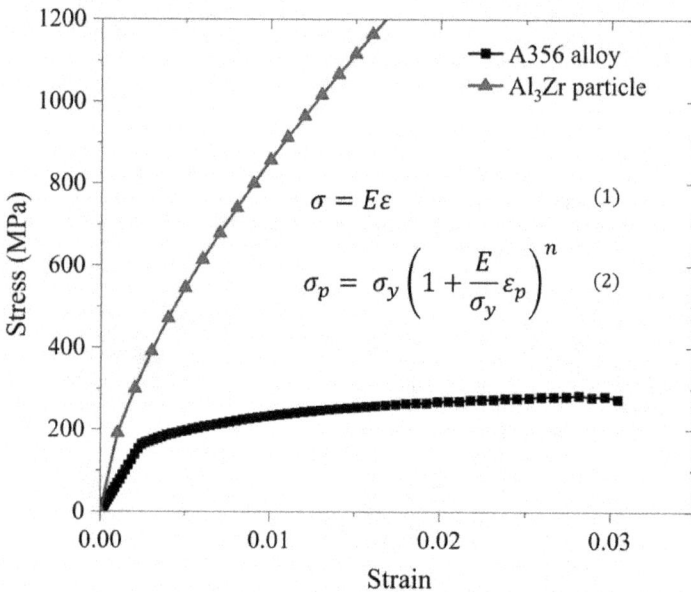

$$\sigma = E\varepsilon \qquad (1)$$

$$\sigma_p = \sigma_y \left(1 + \frac{E}{\sigma_y}\varepsilon_p\right)^n \qquad (2)$$

Figure 1. The stress-strain relationship of A356 alloy and Al₃Zr particle calculated from the reverse algorithm

Materials Research Forum LLC
https://doi.org/10.21741/9781644903094-6

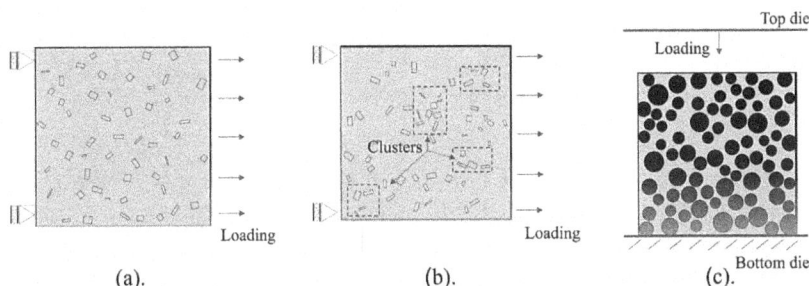

Figure 2. Boundary conditions on (a) uniform distributed RVE, (b) clustered distributed RVE, and (c) foam model

b). Simulation of foam

The structure of foam was generated with a random distribution of pores of 0.8±0.15 mm diameter with 50% porosity. The size of the foam model was a square of length 25mm and meshed with ~ 40000 elements of CPS4R type. The wall thickness (0.08 ±0.012 mm) was varied based on the experimental data. The boundary conditions and the load applied on the foam model are shown in Fig. 2c. Elastic-plastic model was employed for the foam structure in the compression simulation.

Results and discussions

Different sizes of RVE were tested as shown in Fig. 3a and 200 μm with dense mesh (1 μm) with ~ 50,000 elements (CPS4R) was selected to simulate the properties of all microstructural features as well as to minimize the computing time. The effect of particle distribution for 200 μm sized RVE is shown in Fig. 3b, the decrease in stress showed similar behaviour for all the models however the strain values vary because of different particle arrangements, however, the change was minimal. The flow curves of A356/Al$_3$Zr composite with uniform distribution of particles, from the simulation and experimental are shown in Fig. 3c. The simulation results at 5 wt.% Al$_3$Zr showed similar deformation behaviour as of experimental results. The results were slightly overestimated as a consequence of the work hardening model included in the simulation model, which is similar to the work done by Zhou *et al* [24]. The overestimation of the flow curve may also be due to the presence of clusters in the experimental samples, which can hinder the strength of the composite. For the cluster distribution of particles, the stress drop occurred at a low value of strain when compared with the uniform distribution. At higher particle concentrations the distance between the neighbouring particles decreases as a result the change in the failure strain for the cluster and uniform distributed case are similar.

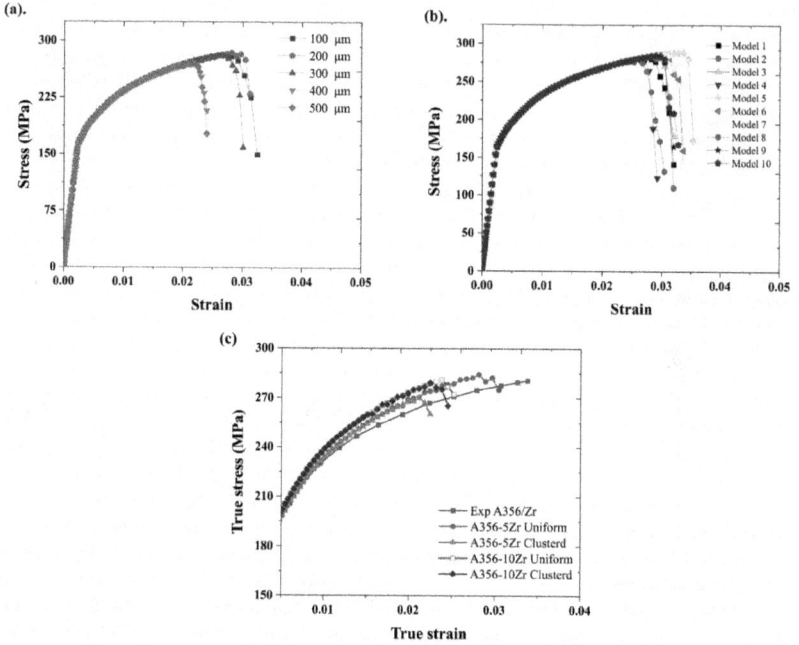

Figure 3. Simulated stress-strain behaviour of A356/Al₃Zr alloy with a). five RVE sizes, b). ten models of 200 μm RVE size, and c) different particle volume fractions and comparison with experimental results.

The Von Mises stress distribution and damage evolution of RVE models with uniform and cluster distribution of *in situ* particles was presented in Fig. 4a and 4b corresponding to the flow curves in Fig. 3c. For the composite with uniform particle distribution the crack initiated within the particle and at the interface with the matrix and propagated at an angle of 45° into the matrix, representing a shear type of fracture (ductile damage) in the matrix. For clustered particle distribution, the cluster served as a stress concentration and both brittle and ductile type of damage evolution is observed. The fracture of the composite occurred at lower strains compared with uniform distribution because of higher stress concentrations and lower tortuosity for crack propagation.

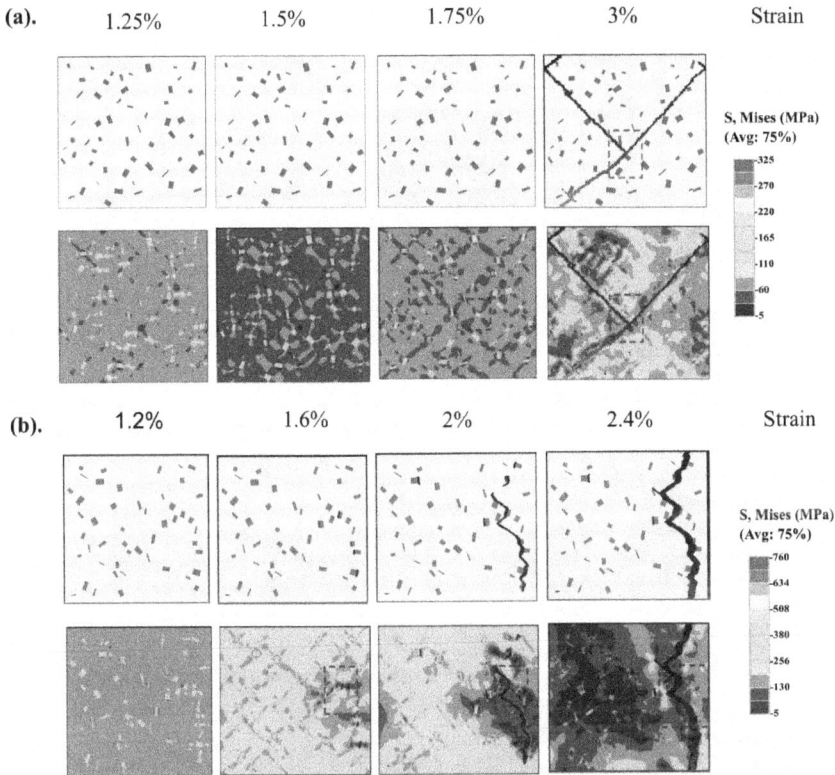

Figure 4. Von Mises stress distribution and damage evolution in A356/Al₃Zr RVE models during tensile deformation in the horizontal direction with a) uniform particle distribution, and (b) clustered particle distribution

The deformation behaviour of A356/Al₃Zr composite metal foam is shown in Fig. 5a, elastic, plateau and densification regions were observed in the simulated results. The deformation bands were formed at 20% strain followed by collapse of the cell wall and densification. The stress-strain behaviour of the composite foam was presented in Fig. 5b. The foam model was validated with pure Al foam and A356/Al₃Zr composite foam with similar properties adopted from previous experimental studies [25, 26]. The experimental and simulation results exhibited similar stress-strain behaviour.

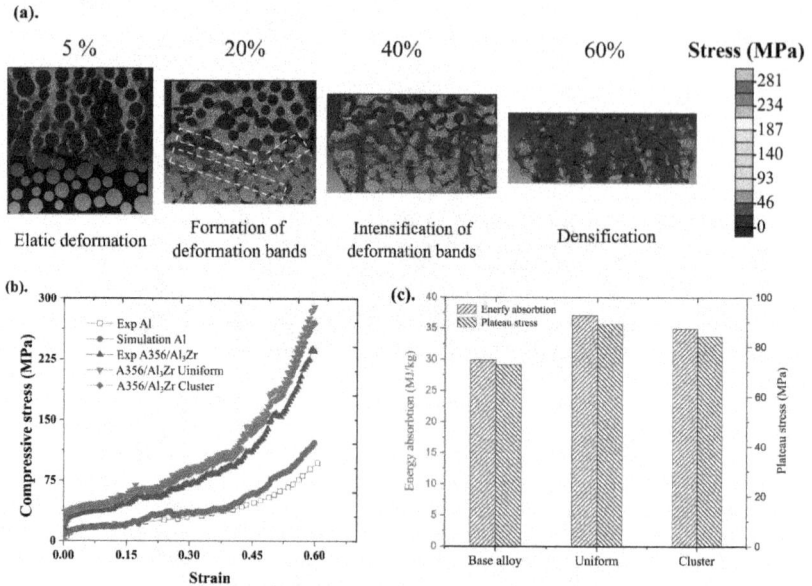

Figure 5. (a). Von Mises stress distribution and deformation behaviour of A356/Al₃Zr foam, and b). compresive stress-strain curves of simulated results and comparison with the experimental data, (c). comparison of plateau stress and energy absorption capacity of foam with uniform and clustered distribution of particles.

The mechanical properties of the composite metal foams were calculated from the simulated stress-strain curves according to ISO 13314 standard [27]. The plateau stress and the energy absorption capacity of the composite foam with the uniform and clustered particle distribution enhanced by 24% and 17% when compared with the base alloy as presented in Fig. 5c. Hence, the composite foam shows better properties when the particles are distributed uniformly within the matrix. The effect of particle shape and size and pore distribution may be studied in future. with two and three-dimensional models of foam to accurately predict the foam properties and optimize the process parameters.

Conclusions

- Numerical simulation modelling of *in situ* foam with uniform and clustered distribution were performed successfully in this study.
- In MMC, the particle fracture was observed followed by initiation and propagation of crack during plastic deformation.
- The simulation results indicated that *in situ* MMCs exhibited improved properties with the use of uniform distribution of particles, this was attributed to lower stress concentration and delay in crack initiation and propagation. However, the mechanical performance of the composite with a cluster distribution with a COV_d of 0.325, and severe stress concentrations lowered the mechanical performance of the composite.
- Similarly, the uniform foam model exhibited better performance due to the enhanced properties of cell wall material.

Porous Metals and Metallic Foams: MetFoam 2023
Materials Research Proceedings 39 (2024) 42-50

Materials Research Forum LLC
https://doi.org/10.21741/9781644903094-6

References

[1] Gibson, L.J., Ashby, M.F.: Cellular Solids. Cell. Solids. (1997). https://doi.org/10.1017/cbo9781139878326

[2] Banhart, J.: Manufacture, characterisation and application of cellular metals and metal foams. Prog. Mater. Sci. 46, 559–632 (2001). https://doi.org/10.1016/S0079-6425(00)00002-5

[3] Banhart, J.: Aluminium foams for lighter vehicles. Int. J. Veh. Des. 37, 114–125 (2005). https://doi.org/10.1504/IJVD.2005.006640

[4] Banhart, J.: Metal foams: Production and stability. Adv. Eng. Mater. 8, 781–794 (2006). https://doi.org/10.1002/adem.200600071

[5] Simone, A.E., Gibson, L.J.: Aluminum foams produced by liquid-state processes. Acta Mater. 46, 3109–3123 (1998). https://doi.org/10.1016/S1359-6454(98)00017-2

[6] Li, D.W., Li, J., Li, T., Sun, T., Zhang, X.M., Yao, G.C.: Preparation and characterization of aluminum foams with ZrH2 as foaming agent. Trans. Nonferrous Met. Soc. China (English Ed. 21, 346–352 (2011). https://doi.org/10.1016/S1003-6326(11)60720-6

[7] Rakesh, M., Srivastava, N., Bhagavath, S., Karagadde, S.: Role of In Situ Formed (Al3Zr + Al3Ti) Particles on Nucleation of Primary Phase in Al-5 wt.% Cu Alloy. J. Mater. Eng. Perform. 2, (2023). https://doi.org/10.1007/s11665-023-08417-z

[8] Atturan, U.A., Nandam, S.H., Murty, B.S., Sankaran, S.: Processing and characterization of in-situ TiB2 stabilized closed cell aluminium alloy composite foams. Mater. Des. 101, 245–253 (2016). https://doi.org/10.1016/j.matdes.2016.03.153

[9] Chawla, N., Sidhu, R.S., Ganesh, V. V.: Three-dimensional visualization and microstructure-based modeling of deformation in particle-reinforced composites. Acta Mater. 54, 1541–1548 (2006). https://doi.org/10.1016/j.actamat.2005.11.027

[10] Ma, S., Wang, X.: Mechanical properties and fracture of in-situ Al3Ti particulate reinforced A356 composites. Mater. Sci. Eng. A. 754, 46–56 (2019). https://doi.org/10.1016/j.msea.2019.03.044

[11] Ma, S., Zhang, X., Chen, T., Wang, X.: Microstructure-based numerical simulation of the mechanical properties and fracture of a Ti-Al3Ti core-shell structured particulate reinforced A356 composite. Mater. Des. 191, 108685 (2020). https://doi.org/10.1016/j.matdes.2020.108685

[12] Antunes, F. V., Ferreira, J.A.M., Capela, C.: Numerical modelling of the Young's modulus of syntactic foams. Finite Elem. Anal. Des. 47, 78–84 (2011). https://doi.org/10.1016/j.finel.2010.09.007

[13] Nammi, S.K., Myler, P., Edwards, G.: Finite element analysis of closed-cell aluminium foam under quasi-static loading. Mater. Des. 31, 712–722 (2010). https://doi.org/10.1016/j.matdes.2009.08.010

[14] Pérez, L., Mercado, R., Alfonso, I.: Young's modulus estimation for CNT reinforced metallic foams obtained using different space holder particles. Compos. Struct. 168, 26–32 (2017). https://doi.org/10.1016/j.compstruct.2017.02.017

[15] Shen, H., Brinson, L.C.: Finite element modeling of porous titanium. Int. J. Solids Struct. 44, 320–335 (2007). https://doi.org/10.1016/j.ijsolstr.2006.04.020

[16] Srivastava, N., Bhagavath, S., Karagadde, S.: Effect of in situ Al3Zr particles on controlling the pore morphology of Al6061 alloy foams. Mater. Today Commun. 26, 101853 (2021). https://doi.org/10.1016/j.mtcomm.2020.101853

[17] An, X., Liu, Y., Ye, J., Li, X.: Fabrication of in-situ TiC–TiB2 reinforced Al foam with enhanced properties, (2020)

[18] Yang, X., Yang, K., Wang, J., Shi, C., He, C., Li, J., Zhao, N.: Compressive Response and Energy Absorption Characteristics of In Situ Grown CNT-Reinforced Al Composite Foams. Adv. Eng. Mater. 19, 294–302 (2017). https://doi.org/10.1002/adem.201700431

[19] Ma, S., Zhuang, X., Wang, X.: Particle distribution-dependent micromechanical simulation on mechanical properties and damage behaviors of particle reinforced metal matrix composites. J. Mater. Sci. 56, 6780–6798 (2021). https://doi.org/10.1007/s10853-020-05684-2

[20] Pandee, P., Sankanit, P., Uthaisangsuk, V.: Structure-mechanical property relationships of in-situ A356/Al3Zr composites. Mater. Sci. Eng. A. 866, (2023). https://doi.org/10.1016/j.msea.2023.144673

[21] Dao, M., Chollacoop, N., Van Vliet, K.J., Venkatesh, T.A., Suresh, S.: COMPUTATIONAL MODELING OF THE FORWARD AND REVERSE PROBLEMS IN INSTRUMENTED SHARP INDENTATION. (2001)

[22] Okereke, M.I., Akpoyomare, A.I.: A virtual framework for prediction of full-field elastic response of unidirectional composites. Comput. Mater. Sci. 70, 82–99 (2013). https://doi.org/10.1016/j.commatsci.2012.12.036

[23] Akpoyomare, A.I., Okereke, M.I., Bingley, M.S.: Virtual testing of composites: Imposing periodic boundary conditions on general finite element meshes. Compos. Struct. 160, 983–994 (2017). https://doi.org/10.1016/j.compstruct.2016.10.114

[24] Zhou, J., Gokhale, A.M., Gurumurthy, A., Bhat, S.P.: Realistic microstructural RVE-based simulations of stress-strain behavior of a dual-phase steel having high martensite volume fraction. Mater. Sci. Eng. A. 630, 107–115 (2015). https://doi.org/10.1016/j.msea.2015.02.017

[25] Yang, X., Hu, Q., Du, J., Song, H., Zou, T., Sha, J., He, C., Zhao, N.: Compression fatigue properties of open-cell aluminum foams fabricated by space-holder method. Int. J. Fatigue. 121, 272–280 (2019). https://doi.org/10.1016/j.ijfatigue.2018.11.008

[26] Bensalem, I., Benhizia, A.: Novel design of irregular closed-cell foams structures based on spherical particle inflation and evaluation of its compressive performance. Thin-Walled Struct. 181, (2022). https://doi.org/10.1016/j.tws.2022.109991

[27] Abstract of : ISO 13314 : 2011 Mechanical testing of metals — Ductility testing — Compression test for porous and cellular metals Foreword 1 Scope 2 Normative references 3 Terms and definitions. 1–3

Porous Metals and Metallic Foams: MetFoam 2023
Materials Research Proceedings 39 (2024) 51-58

Materials Research Forum LLC
https://doi.org/10.21741/9781644903094-7

Opportunities of Metal Structures in Cooling Systems

Mandy Uhlig[1,a] *, Julius Eik Grimmenstein[2,b] Pauline Langbehn[3,c]
and Ralf Döring[4,d]

[1]Fraunhofer IWU, Reichenhainer Str. 88, 09126 Chemnitz, Germany

[2]TU Bergakademie Freiberg, Lampadiusstraße 4, 09599 Freiberg, Germany

[3]iPoint-system GmbH, Max-Brauer-Allee 50, 22765 Hamburg, Germany

[4]Fraunhofer ENAS, Technologie-Campus 3, 09126 Chemnitz, Germany

[a]mandy.uhlig@iwu.fraunhofer.de, [b]Julius-Eik.Grimmenstein@iart.tu-freiberg.de,
[c]pauline.langbehn@ipoint-systems.de, [d]ralf.doering@enas.fraunhofer.de

Keywords: Cellular Metal Structures, Cooling, Power Electronics

Abstract. The growing market of power electronics in the mobility sector leads to an increasing demand for cooling systems. In the project this need to improve performance is to be met by adapting the cooling structure. Depending on the intended application of cooling systems - automotive, railway and aerospace - different requirements are defined for the cooling process resulting in varying conditions for the design. So metallic foam structures are under investigation because of their high inner surface. Two different process lines are most suitable for the aimed application. The production and optimization of galvanized foams seems to be the most lucrative for a low-cost product, while 3D printing is currently only worthwhile for special applications such as aerospace. As a potentially more cost-effective process, which is already being used for small series, investment casting structures are being investigated as an alternative. Depending on the production process chosen, corresponding requirements for structure creation suitable for production apply. Corresponding process adaptations are also taken into consideration. The first optimization step is an analysis of the conventional open cellular metal foam structures using CT. The results of the CT evaluation, together with the empirical data for fluid mechanical and thermal characteristics, are the basis for a later replacement model of the CFD simulation. Besides the Kelvin cell, which is a good geometrical substitute for the conventional structures that copy the Polyurethan master pattern, other cell types are also considered. Alternatively, structures based on mathematical cells e.g. Schwarz P/D, offer the possibility of separating two media to create cross-flow or counter-flow heat exchangers. Regardless of the chosen system, the main task of the investigation is to find an optimum of the relation between pressure drop and heat transfer performance for the corresponding system and to design the cell arrangement in a way that is suitable for manufacturing. This justifies, among other things, the investigation of a minimal surface structure, which at first seems contradictory. Considering the manufacturing process to be defined beforehand, requirements such as self-supporting design (additive process) and free accessibility (electroplating) also play an important role in the structure development. Thus, the goal is still to optimize the cell and web geometry accordingly.

Introduction

For the improvement of commercial cooling systems for power electronics the application of cellular structures seems to be promising in order to enlarge the surface for heat transfer. As already investigated in several publications the use of metal foams can improve the thermal distribution for cooling[1]. To obtain reliable results, the use of cellular structures designed, adapted and evaluated for three well-defined use cases is objective of a current research project. In the KoLibri project, funded by the BMWK, the cooling performance for high-performance electronics in the mobility sector is being investigated. Three use cases were distinguished: the automotive sector, railway industry and the aerospace sector. All of those topics show partly conflicting requirements for the design of (new) cooling systems. In order to be able to meet the corresponding requirement profiles, simulation plays an important role in the design in addition to construction. Furthermore, the differing requirements lead to different manufacturing processes, such as the galvanic or investment casting.

Finally, the eco-balance has to be kept in mind. A not less important goal of the structural design is the consideration of the ecological impact already in the first engineering steps.

Cooling of high-performance electronics in the mobility sector

The division of the mobility sector into three main industries is necessary for a sensible design of a cooling solution, not least because of the different unit numbers, prices and requirements.

Whereas in the aerospace industry, price tends to take second place to safety and weight savings, in the automotive industry it is the driving factor.

For the automotive industry, for example, it is also necessary to consider an alternative manufacturing route, since investment casting, which is sufficient for the low volumes in the aerospace sector, is rather not an option.

One possibility to withstand the cost pressure would be the electroplating process suitable for large series production. This also seems to be a good alternative due to the material component copper. However, new challenges also arise here. For example, wettability and permeability for the electrolyte into the structural design of cells. At this point, simulation is of elementary importance for any product design. In order to create a holistic picture, the thermal-electrical simulation is carried out first.

An essential and permanent part of the simulation setup is the validation based on measurement results. For this purpose, the demonstrator underwent an evolution (Figure 1). It was further developed from a highly geometrically reduced model to a test setup adapted to the PinFin cooler to the final benchmark the commercial PinFin solution.

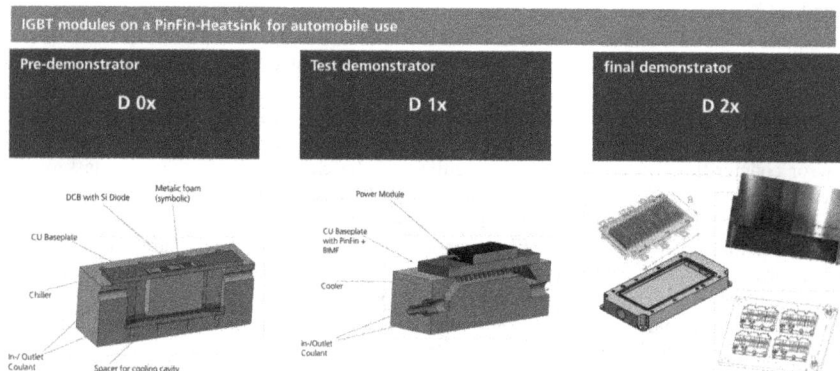

Figure 1   Evolution of the Test demonstrator

In addition to the pure validation, these demonstrators should also support the coupling of the different simulation modules (electro-thermal + fluid dynamic).

As a final step, a link with the LCA software "UMBERTO" is intended to enable optimization at this level as well and to include it at the earliest possible stage.

Evaluation of the thermal behavior of power modules using numerical simulation

Together with the partner ILK, Fraunhofer ENAS developed a methodology for the simulation of the thermal conduction behavior of the overall assembly. This work aims at combining the two fundamentally different simulation applications "electro-thermal evaluation of power electronics" and "fluid dynamic evaluation of cooling". Since the results of both simulations influence each other (a better cooling causes a change of the power dissipation of the power module and vice versa), the coupling of both simulation types via a common interface is necessary and shown on the realized demonstrator in Figure 2.

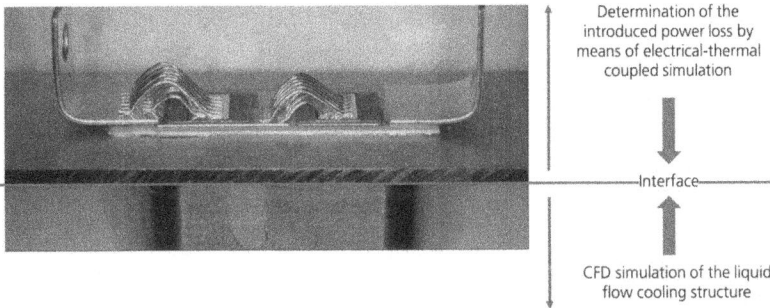

Figure 2   Assembly used to validate the combination of electro-thermal and the CFD simulation

The focus of the work of Fraunhofer ENAS is the development of an electro-thermal simulation model. **In a first step**, this model is calibrated based upon the concept demonstrators, which are manufactured and experimentally evaluated by the partner Siemens. For this purpose, all input parameters necessary for a trustworthy simulation, such as correct geometry and material data, were determined and the model was built. In several iterations, the model was calibrated using experimentally determined data from the partner Siemens (Figure 3).

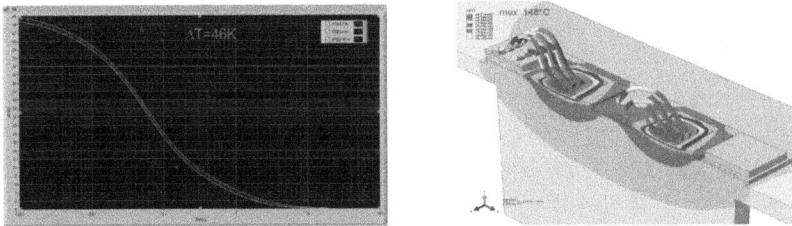

Figure 3   Calibration of the junction temperature of the dies by means of electro-thermal coupled simulation

Furthermore, for the fluid dynamic simulations of the partner ILK, the expected heat transfer conditions are transferred at the "simulation" interface (Figure 4), so that the electro-thermal simulation can serve as boundary condition for the CFD simulation.

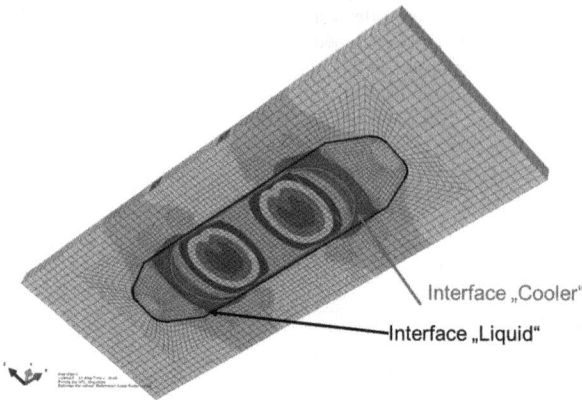

Figure 4   Averaged Heat Flows at the "simulation" interface

For this, however, it is necessary to drastically reduce the effort for the electro-thermal simulation. Therefore, **in the second step**, this model is converted into a simplified behavior model – a so called "surrogate model". By means of a DoE, the surrogate model leads to a response surface (Figure 5).

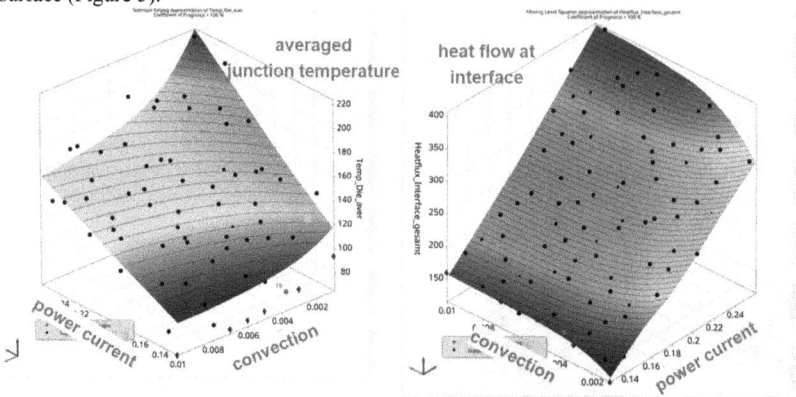

Figure 5   Surrogate Modell (Response Surface Method)

With the help of the surrogate model, results were predicted and further on successfully verified by experimental measurements. The surrogate model is therefore trustworthy.

The next steps include the conversion of the surrogate model into a recognized exchange format. That way it can be used in other simulations as well as integrated into system simulation programs. Furthermore, it is planned to extend the surrogate model in multiple directions such as the consideration of ecological aspects in model design. Another use would be the integration of the thermomechanical reliability of the assembly and connection technology used, represent a major challenge.

Some of the simulations shown above are related to solutions that are partly state of the art as the design of the coolant flow wasn't changed much. To step ahead it was mentioned already that

Porous Metals and Metallic Foams: MetFoam 2023
Materials Research Proceedings 39 (2024) 51-58

Materials Research Forum LLC
https://doi.org/10.21741/9781644903094-7

one approach is to optimize the heat distribution and influence the flow of the cooling liquid by application of cellular metal structures.

Optimizing cellular metal structures to create maximum cooling efficiency

Based on the results of the electro-thermal simulation, the structural optimization can also be started. After a detailed examination of the structure of stochastic foams by means of CT, the influence of the cell structure needs to be evaluated on the basis of different CAD-modelled cells. Thus, the mathematical description is done already in several publications e.g. based upon Kelvin cells.[2]

One intention using those substitute cell models is the optimization adapted to the thermal load case. By using a combination of topology optimization and the Software SYNERA a homogenous cell structure can be changed in strut thickness to the thermal load case represented by a thermographic picture. That way the flow resistance and heat transition can be optimized for any location within the network leading to the best possible cell structure.

Parameter determination for the structure optimization

Some technologies such as galvanic coating and investment casting have been mentioned already. All design and LCA development processes are based upon parameters of the production. In order to get precise figures for those simulation and computer aided optimization it is necessary to produce several samples for testing. The two mentioned processes are taken into account, the galvanic process, as copper foams are already commercially available (6) and the investment casting process shown in 7.

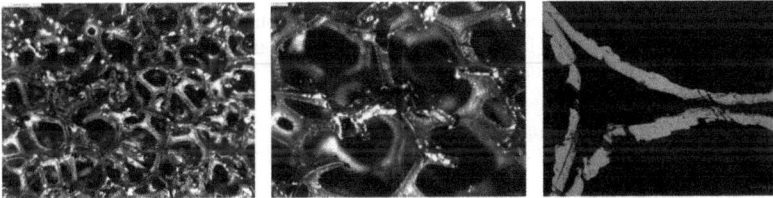

Figure 6   *Commercially available copper foam*

The CAD-structures (a) are 3d-printed in wax (b), which is cheaper than a commercial metal-print and has a resolution down to 16 μm. Afterwards the investment casting process starts with the embedding of that master pattern in molding material (c) a dewaxing and heat treatment of the mold, followed by infiltration of the emerged cavity with molten metal. Once solidified the structure can be unmolded by destroying the "invested" mold (d) and is finally cleaned by high pressure water jet. Finally, a post machining such as cutting the gating and feeding system is resulting in the wanted metal structure (e).

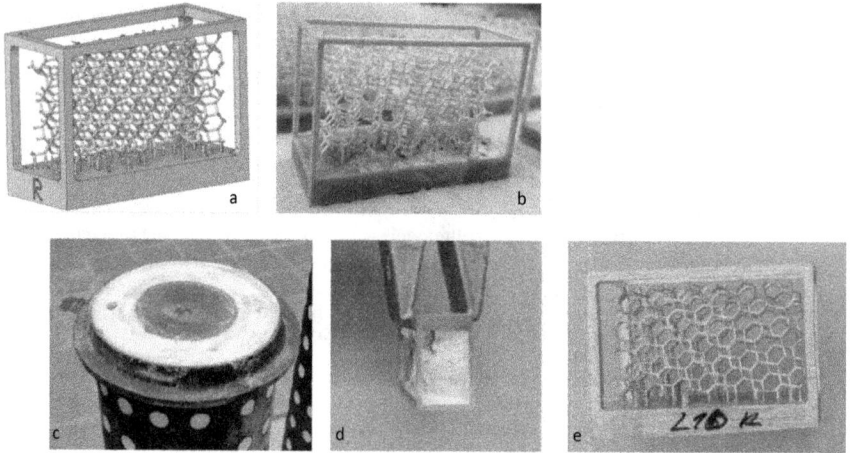

Figure 7   Investment casting process for an open cellular metal structure

Reporting the whole process and parameter determination lead to the next important step of the project.

Life Cycle Assessment (LCA) and recycling of metal foam-based cooling structures
The importance of an LCA for all processes is shown in 8 illustrating the tremendous environmental impact of a product in the "end-of-life" phase.

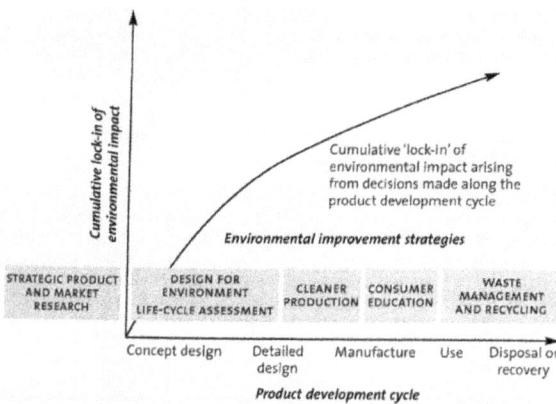

Figure 8   Investment casting process for an open cellular metal structure [5]

For every product in today's, a life cycle analysis and a suitable recycling concept are already in the foreground in the development phase, in order to create a sustainable and holistically considered product and processes right from the start. In this context, LCA is used, among other things, to integrate sustainability aspects during development or to create process routes, including these processes that are downstream from the actual life cycle, such as recycling [3][4].

In order to be able to compile these in detail, well-founded data must be available. When a new process or product is developed, this data is usually not available and data gaps occur which have to be closed by tests and assumptions until final data is determind. In order to provide a first overview, a screening LCA is usually set up, which is not ISO compliant, but provides a first reference point to select suitable processes in the course of a new development [6]. In this context, the LCA of the recycling route is particularly important, as it has a direct impact on the whole product, due to the good LCA of the recirculated materials. These aforementioned data gaps were also present in the development of foam-based coolers, and in order to close them and detect a suitable recycling route, preliminary tests were conducted to investigate different process routes for existing cooling systems in industry as benchmarks. They differ mainly in the previous disassembly and the different shredding machines.

The products mentioned above were studied to investigate a suitable route of the conventional cooling structures. Also to support those investigations with performance data in order to detect a preferred variant, which can then be analogized to the foam-based coolers, or compared with the final process so that the life cycle assessments can be compared. The energy requirements of the process routes were measured using in-machine technology.

Based upon the recorded data, the route with an granulator used for shredding the scrap emerged as the most energy-efficient and thus the most environmentally friendly. It is true that the route with prior disassembly was also promising, as this increased the degree of recycling. However, this was neglected for further investigations, since no foam-based complete structures are yet available and thus no disassembly can take place.

Since, among other things, cooling structures made of coated copper foam -based on polyurethane (PUR)- are to be used and these structures require the greatest development effort, this process was transferred to these metallic foams in a first step. It was found that only 19.41% of these foams were decoated. This means that only about one fifth of the total copper on the surface could be removed and separated. This was done by air separation and density sorting using the float-sink method, which works by adjusting the density of a fluid, causing the copper to sink and the PUR to rise. The exact proportion was determined using the formula:

$$Degree\ of\ decoating = \frac{Mass\ PUR\ in\ light\ fraction}{Mass\ PUR\ in\ Input\ material} \times 100$$

The weight of the PUR in the input material was measured by the supplier before coating and the weight of the PUR in the light fraction, was determined by X-ray fluorescence analysis. It must be assumed that the PUR in the heavy fraction is still bound to the copper, as it cannot be separated out.

Based upon the results of the preliminary tests, a suitable process route must be developed in further investigations that significantly increases the degree of decoating. In addition, the inputs and outputs of the life cycle assessment must be presented in more detail in the future so that no CO2 emitter is omitted.

Summary
In conclusion, the correlating results from simulation and testing already demonstrate improvements in cooling performance by using the stochastic foams used initially. The issue of LCA and recycling needs to be addressed anew for the process to be developed, as commercially available methods do not produce the desired effect.

Overall, the current results are a pointer in the right direction and reveal not only great potential but also a need for action, especially with regard to the implementation of ecological specifications in the initial development stages of a new product. Furthermore, the use of open-cell metal structures for cooling technology, among other applications, is unstoppable, although it is clear that stochastic structures are only the starting point for the development of application-specific structures due to their limited design capability. First approaches to overcome that issue are shown.

References

[1] C.Y. Zhao,Review on thermal transport in high porosity cellular metal foams with open cells, International Journal of Heat and Mass Transfer,Volume 55, Issues 13–14,2012, https://doi.org/10.1016/j.ijheatmasstransfer.2012.03.017

[2] Sir Thomson, W. On the division of space with minimum partitional area. Acta Math. 11, 121–134 (1887). https://doi.org/10.1007/BF02612322

[3] Principles of Environmental Protection Standards Committee (2021): DIN EN ISO 14040 – Environmental management - Life cycle assessment - Principles and framework. DIN German Institute for Standardization

[4] Principles of Environmental Protection Standards Committee (2021): DIN EN ISO 14044 – Environmental management - Life cycle assessment - Requirements and guidelines. DIN German Institute for Standardization

[5] H. Lewis, et al., Design + Environment :A Global Guide to Designing Greener Goods, Greenleaf Publishing, Sheffield UK, 2001

[6] S. Beemsterboer, H. Baumann & H. Wallbaum, Ways to get work done: a review and systematisation of simplification practices in the LCA literature. Int J Life Cycle Assess 25, 2154–2168 (2020). https://doi.org/10.1007/s11367-020-01821-w

Keyword Index

www.ingramcontent.com/pod-product-compliance
Lightning Source LLC
Chambersburg PA
CBHW071514210326
41597CB00018B/2753